Introduction to Engineering Ethics

Second Edition

Mike W. Martin
Professor of Philosophy
Chapman University

Roland Schinzinger
Late Professor Emeritus of Electrical Engineering
University of California, Irvine

**McGraw-Hill
Higher Education**

Boston   Burr Ridge, IL   Dubuque, IA   New York   San Francisco   St. Louis
Bangkok   Bogotá   Caracas   Kuala Lumpur   Lisbon   London   Madrid   Mexico City
Milan   Montreal   New Delhi   Santiago   Seoul   Singapore   Sydney   Taipei   Toronto

McGraw-Hill
Higher Education

INTRODUCTION TO ENGINEERING ETHICS, SECOND EDITION
Published by McGraw-Hill, a business unit of The McGraw-Hill Companies, Inc., 1221 Avenue of the Americas, New York, NY 10020. Copyright © 2010 by The McGraw-Hill Companies, Inc. All rights reserved. Previous edition © 2000. No part of this publication may be reproduced or distributed in any form or by any means, or stored in a database or retrieval system, without the prior written consent of The McGraw-Hill Companies, Inc., including, but not limited to, in any network or other electronic storage or transmission, or broadcast for distance learning.

Some ancillaries, including electronic and print components, may not be available to customers outside the United States.

This book is printed on acid-free paper.

1 2 3 4 5 6 7 8 9 0 DOC/DOC 0 9

ISBN 978-0-07-248311-6
MHID 0-07-248311-3

Global Publisher: *Raghothaman Srinivasan*
Sponsoring Editor: *Debra B. Hash*
Director of Development: *Kristine Tibbetts*
Developmental Editor: *Darlene M. Schueller*
Senior Marketing Manager: *Curt Reynolds*
Project Manager: *Joyce Watters*
Senior Production Supervisor: *Laura Fuller*
Associate Design Coordinator: *Brenda A. Rolwes*
Cover Designer: *Studio Montage, St. Louis, Missouri*
(USE) Cover Image: *Globe Radiating Light, Circuit Boards, Montage of Computer Parts:* © *Getty Images; Pyramid Shaped Tomb:* © *1998 Copyright IMS Communications Ltd./Capstone Design*
Senior Photo Research Coordinator: *John C. Leland*
Compositor: *Lachina Publishing Services*
Typeface: *10/12 Times Roman*
Printer: *R.R. Donnelley Crawfordsville, IN*

Library of Congress Cataloging-in-Publication Data

Martin, Mike W., 1946-
    Introduction to engineering ethics / Mike W. Martin, Roland Schinzinger.—2nd ed.
        p. cm.
Rev. ed. of Introduction to engineering ethics / Roland Schinzinger, Mike W. Martin. 2000.
Includes bibliographical references and index.
ISBN 978-0-07-248311-6—ISBN 0-07-248311-3 (hard copy : alk. paper)
I. Schinzinger, Roland. Introduction to engineering ethics. II. Title.
TA157.S382 2010
174'.962—dc22

                                    2008047516

www.mhhe.com

In memory of Roland Schinzinger—
Inspiring mentor, friend, and advocate for peace.
Mike W. Martin

CONTENTS

ABOUT THE AUTHORS

Mike W. Martin and **Roland Schinzinger** began their 25-year collaboration as a philosopher-engineer team in the National Project on Philosophy and Engineering Ethics, 1978–1980. They have coauthored articles, team-taught courses, and given presentations to audiences of engineers and philosophers. In 1992 they received the Award for Distinguished Literary Contributions Furthering Engineering Professionalism from The Institute of Electrical and Electronics Engineers, United States Activities Board.

Introduction to Engineering Ethics is a condensed and updated version of their book, *Ethics in Engineering,* which has been published in several editions and translations.

Mike W. Martin received his BS and MA from the University of Utah, and his PhD from the University of California, Irvine, and he is currently professor of philosophy at Chapman University. His books include *Creativity: Ethics and Excellence in Science* (2007), *Everyday Morality* (2007), *Albert Schweitzer's Reverence for Life* (2007), *From Morality to Mental Health* (2006), and *Meaningful Work: Rethinking Professional Ethics* (2000). A member of Phi Beta Kappa and Phi Kappa Phi, he received the Arnold L. and Lois S. Graves Award for Teachers in the Humanities, two fellowships from the National Endowment for the Humanities, and several teaching awards from Chapman University.

Roland Schinzinger (1926–2004) received his BS, MS, and PhD in electrical engineering from the University of California at Berkeley, and he was a founding faculty member to the University of California at Irvine. Born and raised in Japan, where he had industrial experience with several companies, he worked in the United States as a design and development engineer at Westinghouse Electric Corporation. He is author or coauthor of *Conformal Mapping: Methods and Applications* (1991, 2003), *Emergencies in Water Delivery* (1979), and *Experiments in Electricity and Magnetism* (1961). His honors include the IEEE Centennial and Third Millennium medals, Fellow of IEEE, and Fellow of AAAS.

PREFACE

Technology has a pervasive and profound effect on the contemporary world, and engineers play a central role in all aspects of technological development. To hold paramount the safety, health, and welfare of the public, engineers must be morally committed and equipped to grapple with ethical dilemmas they confront.

Introduction to Ethics in Engineering provides an introduction to the issues in engineering ethics. It places those issues within a philosophical framework, and it seeks to exhibit their social importance and intellectual challenge. The goal is to stimulate reasoning and to provide the conceptual tools necessary for responsible decision making.

In large measure we proceed by clarifying key concepts, discussing alternative views, and providing relevant case study material. Yet in places we argue for particular positions that in a subject such as ethics can only be controversial. We do so because it better serves our goal of encouraging responsible reasoning than would a mere digest of others' views. We are confident that such reasoning is possible in ethics, and that, through engaged and tolerant dialogue, progress can be made in dealing with what at first seem irresolvable difficulties.

Second Edition

The book has expanded from 6 to 10 chapters. In addition to new case studies such as global warming and Hurricane Katrina, increased coverage is given to moral reasoning and codes of ethics, personal commitments in engineering, environmental ethics, honesty and research integrity, the philosophy of technology, and peace engineering. "Micro issues" concerning choices by individuals and corporations are connected throughout the book with "macro issues" about broader social concerns.

Case studies appear throughout the text, frequently as part of the Discussion Topics. Those cases not described in great detail offer the opportunity for practice in literature searches. Most of our case studies are based on secondary sources. Thus, each case carries with it an *implied* statement of the sort "If engineer X and company Y did indeed act in the way described, then. . . ." It is important to avoid inflexible conclusions regarding persons or organizations based on one or two cases from the past. Persons can and do change with time for the better or for the worse, and so can organizations.

Basic Engineering Series and Tools

McGraw-Hill's BEST—Basic Engineering Series and Tools—consists of modularized textbooks and applications appropriate for the topic covered in most introductory engineering courses. The goal of the series is to provide the educational community with material that is timely, affordable, of high quality, and flexible in how it is used. For a list of BEST titles, visit our website at www.mhhe.com/engcs/general/best.

Electronic Textbook Options

This text is offered through CourseSmart for both instructors and students. CourseSmart is an online browser where students can purchase access to this and other McGraw-Hill textbooks in digital format. Through their browser, students can access the complete text online at almost half the cost of a traditional text. Purchasing the eTextbook also allows students to take advantage of CourseSmart's web tools for learning, which include full text search, notes and highlighting, and email tools for sharing notes between classmates. To learn more about CourseSmart options, contact your sales representative or visit www.CourseSmart.com.

ACKNOWLEDGMENTS

Many individuals have influenced our thinking about engineering ethics. We wish to thank especially Robert J. Baum, Michael Davis, Dave Dorchester, Walter Elden, Charles B. Fleddermann, Albert Flores, Alastair S. Gunn, Charles E. (Ed) Harris, Joseph R. Herkert, Jacqueline A. Hynes, Deborah G. Johnson, Ron Kline, Edwin T. Layton, Jerome Lederer, Heinz C. Luegenbiehl, Carl Mitcham, Steve Nichols, Kevin M. Passino, Michael J. Rabins, Jimmy Smith, Michael S. Pritchard, Harold Sjursen, Carl M. Skooglund, John Stupar, Stephen H. Unger, Pennington Vann, P. Aarne Vesilind, Vivien Weil, Caroline Whitbeck, and Joseph Wujek.

We also thank the reviewers who provided many helpful suggestions in developing this edition.

Rosalyn W. Berne
University of Virginia

Nicole Larson
Western Washington University

Donald G. Lemke
University of Illinois at Chicago

Gene Moriarty
San Jose State University

David L. Prentiss
University of Rhode Island

R. Keith Stanfill
University of Florida

Charles F. Yokomoto,
Professor Emeritus
Indiana University-Purdue University Indianapolis

And we thank the many authors and publishers who granted us permission to use copyrighted material as acknowledged in the notes, and also the National Society of Professional Engineers® which allowed us to print its code of ethics in the Appendix.

Our deepest gratitude is to our families, whose love and insights have so deeply enriched our work and our lives.

Mike W. Martin and Roland Schinzinger

CHAPTER 1

Ethics and Professionalism

Engineers create products and processes to improve food production, shelter, energy, communication, transportation, health, and protection against natural calamities—and to enhance the convenience and beauty of our everyday lives. They make possible spectacular human triumphs once only dreamed of in myth and science fiction. Almost a century and a half ago in *From the Earth to the Moon*, Jules Verne imagined American space travelers being launched from Florida, circling the moon, and returning to splash down in the Pacific Ocean. In December 1968, three astronauts aboard an Apollo spacecraft did exactly that. Seven months later, on July 20, 1969, Neil Armstrong took the first human steps on the moon (Figure 1-1). This extraordinary event was shared with millions of earthbound people watching the live broadcast on television. Engineering had transformed our sense of connection with the cosmos and even fostered dreams of routine space travel for ordinary citizens.

Historicus, Inc./RF

Figure 1-1
Neil Armstrong on Moon

Most technology, however, has double implications: As it creates benefits, it raises new moral challenges. Just as exploration of the moon and planets stand as engineering triumphs, so the explosions of the space shuttles, *Challenger* in 1986 and *Columbia* in 2003, were tragedies that could have been prevented had urgent warnings voiced by experienced engineers been heeded. We will examine these and other cases of human error, for in considering ethics and engineering alike we can learn from seeing how things go wrong.

In doing so, however, we should avoid allowing technological risks to overshadow technological benefits. Ethics involves appreciating the vast positive dimensions of engineering that so deeply enrich our lives. To cite only a few examples, each of us benefits from the top 20 engineering achievements of the twentieth century, as identified by the National Academy of Engineering: electrification, automobiles, airplanes, water supply and distribution, electronics, radio and television, agricultural mechanization, computers, telephones, air-conditioning and refrigeration, highways, spacecrafts, Internet, imaging technologies in medicine and elsewhere, household appliances, health technologies, petrochemical technologies, laser and fiber optics, nuclear technologies, and high-performance materials.[1]

This chapter identifies some of the moral complexity in engineering, defines engineering ethics, and states the goals in studying it. It also underscores the importance of accepting and sharing moral responsibility within the corporate setting in which today most engineering takes place, and also the need for a basic congruence between the goals of responsible professionals, professions, and corporations.

1.1 Ethics and Excellence in Engineering

Moral values are embedded in engineering projects as standards of excellence, not "tacked on" as external burdens. This is true of even the simplest engineering projects, as illustrated by the following assignment given to students in a freshman engineering course: "Design a chicken coop that would increase egg and chicken production, using materials that were readily available and maintainable by local workers [at a Mayan cooperative in Guatemala]. The end users were to be the women of a weaving cooperative who wanted to increase the protein in their children's diet in ways that are consistent with their traditional diet, while not appreciably distracting from their weaving."[2]

---

[1] National Academy of Engineering, www.greatachievements.org (accessed October 14, 2008).

[2] Clive L. Dym and Patrick Little, *Engineering Design: A Project-Based Introduction*, 2nd ed. (New York: John Wiley & Sons, 2004), 70.

The task proved more complex than first appeared. Students had to identify feasible building materials, decide between cages or one large fenced area, and design structures for strength and endurance. They had to create safe access for the villagers, including ample head and shoulder room at entrances and a safe floor for bare feet. They had to ensure humane conditions for the chickens, including adequate space and ventilation, comfort during climate changes, convenient delivery of food and water, and protection from local predators that could dig under fences. They also had to improve cleaning procedures to minimize damage to the environment while recycling chicken droppings as fertilizers. The primary goal, however, was to double current chicken and egg production. A number of design concepts were explored before a variation of a fenced-in concept proved preferable to a set of cages. Additional modifications needed to be made as students worked with villagers to implement the design in ways that best served their needs and interests.

In combining myriad design goals and constraints, engineering projects integrate multiple moral values connected with those goals and constraints—for example, safety, efficiency, respect for persons, and respect for the environment. As elsewhere, moral values are myriad, and they can give rise to *ethical dilemmas:* situations in which moral reasons come into conflict, or in which the applications of moral values are problematic, and it is not immediately obvious what should be done. The moral reasons might be obligations, rights, goods, ideals, or other moral considerations. For example, at what point does the aim of increasing chicken and egg production compromise humane conditions for the animals?

Technical skill and morally good judgment need to go together in solving ethical dilemmas, and, in general, in making moral choices. So do competence and conscientiousness, creativity and good character. These combinations were identified by the ancient Greeks, whose word *arete* translates into English as "excellence" or as "virtue." In engineering, as in other professions, excellence and ethics go together—for the most part and in the long run.

Micro and Macro Issues

Today, engineers are increasingly asked to understand excellence and ethics in terms of broader societal and environmental concerns. They need to be prepared to grapple with both micro and macro issues. Micro issues concern the decisions made by individuals and companies in pursuing their projects. Macro issues concern more global issues, such as the directions in technological development, the laws that should or should not be passed, and the collective responsibilities of groups such as engineering professional societies and consumer groups. Both micro and macro issues are important in engineering ethics, and often they are interwoven.[3]

As an illustration, consider debates about sport utility vehicles (SUVs). Micro issues arose concerning the Ford Explorer and also Bridgestone/Firestone, who provided tires for the Explorer. During the late 1990s, reports began to multiply about the tread on Explorer tires separating from the rest of the tire, leading to blowouts and rollovers. By 2002, estimates were that 300 people had died, and another 1,000 people were injured, and more recent estimates place the numbers much higher. Ford and Bridgestone/ Firestone blamed each other for the problem, leading to the breakup of a century-old business partnership. As it turned out, the hazard had multiple sources. Bridgestone/Firestone used a flawed tire design and poor quality control at a major manufacturing facility. Ford chose tires with a poor safety margin, relied on drivers to maintain proper inflation within a very narrow range, and then dragged its feet in admitting the problem and recalling dangerous tires.

In contrast, macro issues center on charges that SUVs are among the most harmful vehicles on the road, especially given their numbers. The problems are many: gas-guzzling, excessive polluting, instability because their height leads to rollovers, greater "kill rate" of other drivers during accidents, reducing the vision of drivers in shorter cars behind them on freeways, and blinding other drivers' vision because of high-set lights. Keith Bradsher estimates that SUVs are causing approximately 3,000 deaths in excess of what cars would have caused: "Roughly 1,000 extra deaths occur each year in SUVs that roll over, compared with the expected rollover death rate if these motorists had been driving cars. About 1,000 more people die each year in cars hit by SUVs than would occur if the cars had been hit by other cars. And up to 1,000 additional people succumb each year to respiratory problems because of the extra smog caused by SUVs."[4] Bradsher believes these numbers will continue to increase as more SUVs are added to the road each year and as older vehicles are resold to younger and more dangerous drivers.

Should "the SUV issue" be examined within engineering as a whole, or at least by representative professional and technical societies? If so, what should be done? Or, in a democratic and capitalistic society, should engineers play a role only as indi-

---

[3] Joseph R. Herkert, "Future Directions in Engineering Ethics Research: Microethics, Macroethics and the Role of Professional Societies," *Science and Engineering Ethics* 7 (2001): 403–14.

[4] Keith Bradsher, *High and Mighty* (New York: PublicAffairs, 2002), xvii– xviii; see also p. 305.

viduals but not as organized groups? Should engineers remain uninvolved, leaving the issue entirely to consumer groups and lawmakers? We leave these questions as discussion questions at the end of the section. Even larger macro issues surround public transportation issues in relation to all automobiles and SUVs as we look to the future with a dramatically increasing population, a shrinking of traditional resources, and concerns about global warming.

Dimensions of Engineering

Let us gain a more detailed understanding of moral complexity in engineering as a product develops from a mental concept to physical completion. Engineers encounter both moral and technical problems concerning variability in the materials available to them, the quality of work by coworkers at all levels, pressures imposed by time and the whims of the marketplace, and relationships of authority within corporations. Figure 1–2 charts the sequence of tasks that leads from the concept of a product to its design, manufacture, sale, use, and ultimate disposal.

The idea of a new product is first captured in a conceptual design, which will lead to establishing performance specifications and conducting a preliminary analysis based on the functional relationships among design variables. These activities lead to a more detailed analysis, possibly assisted by computer simulations and physical models or prototypes. The end product of the design task will be detailed specifications and shop drawings for all components.

Manufacturing is the next major task. It involves scheduling and carrying out the tasks of purchasing materials and components, fabricating parts and subassemblies, and finally assembling and performance-testing the product.

Selling comes next, or delivery if the product is the result of a prior contract. Thereafter, either the manufacturer's or the customer's engineers perform installation, personnel training, maintenance, repair, and ultimately recycling or disposal.

Seldom is the process carried out in such a smooth, continuous fashion as indicated by the arrows progressing down the middle of Figure 1–2. Instead of this uninterrupted sequence, intermediate results during or at the end of each stage often require backtracking to make modifications in the design developed thus far. Errors need to be detected and corrected. Changes may be needed to improve product performance or to meet cost and time constraints. An altogether different, alternative design might have to be considered. In the words of Herbert Simon, "Design is usually the kind of problem solving we call ill-structured . . . you don't start off with a well-defined goal. Nor do you start off with a clear set of alternatives, or perhaps any alternatives at

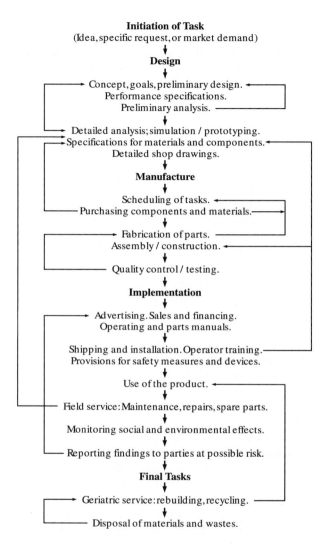

Figure 1–2
Progression of
engineering tasks
(→ ideal progression,
— typical iterations)

all. Goals and alternatives have to emerge through the design
process itself: One of its first tasks is to clarify goals and to begin
to generate alternatives."[5]

This results in an iterative process, with some of the possible
recursive steps indicated by the thin lines and arrows on either

---

[5] Herbert A. Simon, "What We Know about Learning," *Journal of
Engineering Education* (American Society of Engineering Education) 87
(October 1998): 343–48.

side of Figure 1–2. As shown, engineers are usually forced to stop during an initial attempt at a solution when they hit a snag or think of a better approach. They will then return to an earlier stage with changes in mind. Such reconsiderations of earlier tasks do not necessarily start and end at the same respective stages during subsequent passes through design, manufacture, and implementation. That is because the retracing is governed by the latest findings from current experiments, tempered by the outcome of earlier iterations and experience with similar product designs.

Changes made during one stage will not only affect subsequent stages but might also require a reassessment of prior decisions. Dealing with this complexity requires close cooperation among the engineers of many different departments and disciplines such as chemical, civil, electrical, industrial, and mechanical engineering. It is not uncommon for engineering organizations to suffer from "silo mentality," which makes engineers disregard or denigrate the work carried out by groups other than their own. It can be difficult to improve a design or even to rectify mistakes under such circumstances. Engineers do well to establish contact with colleagues across such artificial boundaries so that information can be exchanged more freely. Such contacts become especially important in tackling morally complex problems.

Potential Moral Problems

To repeat, engineering generally does not consist of completing designs or processes one after another in a straightforward progression of isolated tasks. Instead, it involves a trial-and-error process with backtracking based on decisions made after examining results obtained along the way. The design iterations resemble feedback loops, and like any well-functioning feedback control system, engineering takes into account natural and social environments that affect the product and people using it.[6] Let us therefore revisit the engineering tasks, this time as listed in Table 1–1, along with examples of problems that might arise.

The grab bag of problems in Table 1–1 can arise from shortcomings on the part of engineers, their supervisors, vendors, or the operators of the product. The underlying causes can have different forms:

---

[6] Roland Schinzinger, "Ethics on the Feedback Loop," *Control Engineering Practice* 6 (1998): 239–45.

1. *Lack of vision*, which in the form of tunnel vision biased toward traditional pursuits overlooks suitable alternatives, and in the form of groupthink promotes acceptance at the expense of critical thinking.[7]
2. *Incompetence* among engineers carrying out technical tasks.
3. *Lack of time* or *lack of proper materials*, both ascribable to poor management.
4. A *silo mentality* that keeps information compartmentalized rather than shared across different departments.
5. The notion that there are safety engineers *somewhere down the line* to catch potential problems.
6. *Improper use or disposal of the product* by an unwary owner or user.
7. *Dishonesty* in any activity shown in Figure 1–2 and pressure by management to take shortcuts.
8. *Inattention* to how the product is performing after it is sold and when in use.

Although this list is not complete, it hints at the range of problems that can generate moral challenges for engineers. It also suggests why engineers need foresight and caution, especially in imagining who might be affected indirectly by their products and by their decisions, in good or harmful ways.

What Is Engineering Ethics?

In light of this overview of moral complexity in engineering, we can now define engineering ethics and state the goals in studying it. The word *ethics* has several meanings, and hence so does *engineering ethics*. In one sense, ethics is synonymous with morality. It refers to moral values that are sound or reasonable, actions or policies that are morally required (right), morally permissible (all right), or otherwise morally desirable (good). Accordingly, *engineering ethics consists of the responsibilities and rights that ought to be endorsed by those engaged in engineering, and also of desirable ideals and personal commitments in engineering.*

In a second sense, the one used in the title of this book, ethics is the activity (and field) of studying morality; it is an inquiry into ethics in the first sense. It studies which actions, goals, principles, policies, and laws are morally justified. Using this sense, *engineering ethics is the study of the decisions, policies, and values that are morally desirable in engineering practice and research.*

---

[7] Irving Janis, *Groupthink*, 2nd ed. (Boston: Houghton Mifflin, 1982).

Table 1–1 **Engineering tasks and possible problems**

Tasks	A selection of possible problems
Conceptual design	Blind to new concepts. Violation of patents or trade secrets. Product to be used illegally.
Goals; performance specifications	Unrealistic assumptions. Design depends on unavailable or untested materials.
Preliminary analysis	Uneven: Overly detailed in designer's area of expertise, marginal elsewhere.
Detailed analysis	Uncritical use of handbook data and computer programs based on unidentified methodologies.
Simulation, prototyping	Testing of prototype done only under most favorable conditions or not completed.
Design specifications	Too tight for adjustments during manufacture and use. Design changes not carefully checked.
Scheduling of tasks	Promise of unrealistic completion date based on insufficient allowance for unexpected events.
Purchasing	Specifications written to favor one vendor. Bribes, kickbacks. Inadequate testing of purchased parts.
Fabrication of parts	Variable quality of materials and workmanship. Bogus materials and components not detected.
Assembly/ construction	Workplace safety. Disregard of repetitive-motion stress on workers. Poor control of toxic wastes.
Quality control/testing	Not independent, but controlled by production manager. Hence, tests rushed or results falsified.
Advertising and sales	False advertising (availability, quality). Product oversold beyond client's needs or means.
Shipping, installation, training	Product too large to ship by land. Installation and training subcontracted out, inadequately supervised.
Safety measures and devices	Reliance on overly complex, failure-prone safety devices. Lack of a simple "safety exit."
Use	Used inappropriately or for illegal applications. Overloaded. Operations manuals not ready.
Maintenance, parts, repairs	Inadequate supply of spare parts. Hesitation to recall the product when found to be faulty.
Monitoring effects of product	No formal procedure for following life cycle of product, its effects on society and environment.
Recycling/disposal	Lack of attention to ultimate dismantling, disposal of product, public notification of hazards.

These two senses are *normative:* They refer to justified values, desirable (not merely desired) choices, and sound policies. Normative senses differ from *descriptive* senses of ethics. In one descriptive sense, we speak of Henry Ford's ethics, or the ethics of American engineers, referring thereby to what specific individuals or groups believe and how they act, without implying that their beliefs and actions are justified. In another descriptive sense, social scientists study ethics when they describe and explain what people believe and how they act; they conduct opinion polls, observe behavior, examine documents written by professional societies, and uncover the social forces shaping engineering ethics.

In its normative senses, "engineering ethics" refers to justified moral values in engineering, but what are moral values? What is morality? Dictionaries tell us that morality is about right and wrong, good and bad, values and what ought to be done. But such definitions are incomplete, for these words also have nonmoral meanings. Thus, to start a car a person *ought* to put the key in the ignition; that is the *right* thing to do. Again, chocolate tastes *good*, and beauty is an aesthetic *value*. In contrast, morality concerns moral right and wrong, moral good and bad, moral values, and what morally ought to be done. Saying this is not especially illuminating, however, for it is a *circular definition* that uses the word we are trying to define.

As it turns out, morality is not easy to define in any comprehensive way. Of course, we can all give examples of moral values, such as honesty, courage, compassion, and justice. Yet, the moment we try to provide a comprehensive definition of morality we are drawn into at least rudimentary ethical theory. For example, if we say that morality consists in promoting the most good, we are invoking an ethical theory called utilitarianism. If we say that morality is about human rights, we invoke rights ethics. And if we say that morality is essentially about good character, we might be invoking virtue ethics. These and other ethical theories are discussed in Chapter 3.

Why Study Engineering Ethics?

Engineering ethics should be studied because it is *important*, both in contributing to safe and useful technological products and in giving meaning to engineers' endeavors. It is also *complex*, in ways that call for serious reflection throughout a career, beginning with earning a degree. But beyond these general observations, what specific aims should guide the study of engineering ethics?

In our view, the direct aim is to increase our ability to deal effectively with moral complexity in engineering. Accordingly, the study of engineering ethics strengthens our ability to reason clearly and carefully about moral questions. To invoke terms widely used in ethics, the unifying goal is to increase moral autonomy.

Autonomy means *self-determining*, but not just any kind of independent reflection about ethics amounts to moral autonomy. Moral autonomy can be viewed as the skill and habit of thinking rationally about ethical issues on the basis of moral concern and commitment. This foundation of general responsiveness to moral values derives primarily from the training we receive as children in being sensitive to the needs and rights of others, as well as of ourselves. When such training is absent, as it often is with seriously abused children, the tragic result can be an adult sociopath who lacks any sense of moral right and wrong. Sociopaths (or psychopaths) are not morally autonomous, regardless of how independent their intellectual reasoning about ethics might be.

Improving the ability to reflect carefully on moral issues can be accomplished by improving various practical skills that will help produce autonomous thought about moral issues. As related to engineering ethics, these skills include the following.

1. Moral awareness: Proficiency in recognizing moral problems and issues in engineering
2. Cogent moral reasoning: Comprehending, clarifying, and assessing arguments on opposing sides of moral issues
3. Moral coherence: Forming consistent and comprehensive viewpoints based on consideration of relevant facts
4. Moral imagination: Discerning alternative responses to moral issues and finding creative solutions for practical difficulties
5. Moral communication: Precision in the use of a common ethical language, a skill needed to express and support one's moral views adequately to others

These are the *direct* goals in college courses. They center on cognitive skills—skills of the intellect in thinking clearly and cogently. It is possible, however, to have these skills and yet not act in morally responsible ways. Should we therefore add to our list of goals the following goals that specify aspects of moral commitment and responsible conduct?

6. Moral reasonableness: The willingness and ability to be morally reasonable
7. Respect for persons: Genuine concern for the well-being of others as well as oneself
8. Tolerance of diversity: Within a broad range, respect for ethnic and religious differences and acceptance of reasonable differences in moral perspectives
9. Moral hope: Enriched appreciation of the possibilities of using rational dialogue in resolving moral conflicts
10. Integrity: Maintaining moral integrity and integrating one's professional life and personal convictions

In our view we should add these goals to the study of engineering ethics, for without them there would be little practical point in studying ethics. At the same time, the goals are often best pursued implicitly and indirectly, more in how material is studied and taught than in preaching and testing. A foundation of moral concern must be presupposed, as well as evoked and expanded, in studying ethics at the college level.

Discussion Questions

1. Identify the moral values, issues, and dilemmas, if any, involved in the following cases, and explain why you consider them moral values and dilemmas.

   a. An engineer notified his firm that for a relatively minor cost a flashlight could be made to last several years longer by using a more reliable bulb. The firm decides that it would be in its interests not to use the new bulb, both to keep costs lower and to have the added advantage of "built-in obsolescence" so that consumers would need to purchase new flashlights more often.

   b. A linear electron accelerator for therapeutic use was built as a dual-mode system that could either produce X-rays or electron beams. It had been in successful use for some time, but every now and then some patients received high overdoses, resulting in painful after-effects and several deaths. One patient on a repeat visit experienced great pain, but the remotely located operator was unaware of any problem because of lack of communication between them: The intercom was broken, and the video monitor had been unplugged. There also was no way for the patient to exit the examination chamber without help from the outside, and hence the hospital was partly at fault. On cursory examination of the machine, the manufacturer insisted that the computerized and automatic control system could not possibly have malfunctioned and that no one should spread unproven and potentially libelous information about the design. It was the painstaking, day-and-night effort of the hospital's physicist that finally traced the problem to a software error introduced by the manufacturer's efforts to make the machine more user-friendly.[8]

2. Regarding the following example, comment on why you think simple human contact made such a large difference. What does it say about what motivated the engineers, both before and after the encounter? Is the case too unique to permit generalizations to other engineering products?

---

[8] N. B. Leveson and C. Turner, "An Investigation of the Therac-25 Accidents," *Computer* (IEEE, July 1993), 18–41.

A team of engineers are redesigning an artificial lung marketed by their company. They are working in a highly competitive market, with long hours and high stress. The engineers have little or no contact with the firm's customers, and they are focused on technical problems, not people. It occurs to the project engineer to invite recipients of artificial lungs and their families to the plant to talk about how their lives were affected by the artificial lung. The change is immediate and striking: "When families began to bring in their children who for the first time could breathe freely, relax, learn, and enjoy life because of the firm's product, it came as a revelation. The workers were energized by concrete evidence that their efforts really did improve people's lives, and the morale of the workplace was given a great lift."[9]

3. Should SUV problems at the macro level be of concern to engineers as a group and their professional societies? Should individual automotive engineers, in their daily work, be concerned about the general social and environmental impacts of SUVs?

4. It is not easy to define morality in a simple way, but it does not follow that morality is a hopelessly vague notion. For a long time, philosophers thought that an adequate definition of any idea would specify a set of logically necessary and sufficient conditions for applying the idea. For example, each of the following features is logically necessary for a triangle, and together they are sufficient: a plane figure, having three straight lines, closed to form three angles. The philosopher Ludwig Wittgenstein (1889–1951), however, argued that most ordinary (nontechnical) ideas cannot be neatly defined in this way. Instead, there are often only "family resemblances" among the things to which words are applied, analogous to the partly overlapping similarities among members of a family—similar eye color, shape of nose, body build, temperament, and so forth.[10] Thus, a book might be hardback, paperback, or electronic; printed or handwritten; in English or German; and so forth. Can you specify necessary and sufficient conditions for the following ideas: chairs, buildings, energy, safety, engineers, morality?

5. Mention of ethics sometimes evokes groans, rather than engagement, because it brings to mind onerous constraints and unpleasant disagreements. Worse, it evokes images of self-righteousness, hypocrisy, and excessively punitive attitudes of blame and punishment—attitudes that are themselves subject to moral critique. Think of a recent event that led to a public outcry. With regard to the event, discuss the difference between being

---

[9] Mihaly Csikszentmihalyi, *Good Business* (New York: Viking, 2003), 206.
[10] Ludwig Wittgenstein, *Philosophical Investigations*, 3rd ed., trans. G. E. M. Anscombe (New York: Macmillan, 1958), 32.

morally reasonable and being moralistic in a pejorative sense. In doing so, consider such things as breadth of vision, tolerance, sensitivity to context, and commitment.

1.2 Responsible Professionals, Professions, and Corporations

Moral responsibility is an idea that applies to individual engineers, groups of engineers, and the corporations in which most engineers do their work. It is also a multifaceted idea that combines obligations, ideals of character, accountability, praiseworthiness, and blameworthiness. Let us begin with an example of a responsible engineer.

Saving Citicorp Tower

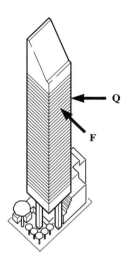

Structural engineer Bill LeMessurier (pronounced "LeMeasure") and architect Hugh Stubbins faced a challenge when they worked on the plans for New York's fifth highest skyscraper. St. Peter's Lutheran Church owned and occupied a corner of the lot designated in its entirety as the site for the new structure. An agreement was reached: The bank tower would rise from nine-story-high stilts positioned at the center of each side of the tower, and the church would be offered a brand new St. Peter's standing freely underneath one of the cantilevered corners. Completed in 1977, the Citicorp Center appears as shown in Figure 1–3. The new church building is seen below the lower left corner of the raised tower.

LeMessurier's structure departed from the usual in that the massive stilts are not situated at the corners of the building, and half of its gravity load as well all of its wind load is brought down an imaginatively designed trussed frame, which incorporates wind braces on the outside of the tower.[11] In addition, LeMessurier installed a tuned mass damper, the first of its kind in a tall building, to keep the building from swaying in the wind.

Questions asked by an engineering student a year after the tower's completion prompted LeMessurier to review certain structural aspects of the tower and pose some questions of his own.[12] For instance, could the structure withstand certain loads caused by strong quartering winds? In such cases, two sides of the building receive the oblique force of the wind, and the resultant force is 40 percent larger than when the wind hits only one face of the structure straight on. The only requirement stated in

FIGURE 1–3
Axonometric view of Citicorp tower with the church in the lower left-hand corner. Wind loads: F, frontal and Q, Quartering. (Adaptation of an axonometric drawing by Henry Dong, Anspach Grossman Portugal, Inc., in Buildings Type Study 492, *Architectural Record*, Mid-August Special Issue [1976]: 66.)

[11] Buildings Type Study 492, Engineering for Architecture: "Citicorp Center and St. Peter's Lutheran Church," *Architectural Record* (Mid-August Special Issue, 1976): 61–71; Charles Thornton, "Conversation with William LeMessurier," in C. H. Thornton et al., *Exposed Structures in Building Design* (New York: McGraw-Hill, 1993).

[12] Joe Morgenstern, "The Fifty-Nine Story Crisis," *New Yorker* (May 29, 1995): 45–53. Check also the Online Ethics Center for Engineering and Science (http://www.onlineethics.org/).

the building code specified adequacy to withstand certain perpendicular wind loads, and that was the basis for the design of the wind braces. But there was no need to worry because the braces as designed could handle such an excess load without difficulty, provided the welds were of the expected high quality.

Nevertheless, the student's questions prompted LeMessurier to place a call from his Cambridge, Massachusetts, office to his New York office, to ask Stanley Goldstein, his engineer in charge of the tower erection, how the welded joints of the bracing structure had worked out. How difficult was the job? How good was the workmanship? To his dismay, Goldstein answered, "Oh, didn't you know? [The joints] were never welded at all because Bethlehem Steel came to us and said they didn't think we needed to do it." The New York office, as it was allowed to do, had approved the proposal that the joints be bolted instead. But again the diagonal winds had not been taken into account.

At first, LeMessurier was not too concerned; after all, the tuned mass damper would still take care of the sway. So he turned to his consultant on the behavior of high buildings in wind, Alan Davenport at the University of Western Ontario. On reviewing the results of his earlier wind tunnel tests on a scaled-down Citicorp Center, Davenport reported that a diagonal wind load would exceed the perpendicular wind load by much more than the 40 percent increase in stress predicted by an idealized mathematical model. Winds sufficient to cause failure of certain critical bolted joints—and therefore of the building—could occur in New York every sixteen years. Fortunately, those braces that required strengthening were accessible, but the job would be disruptive and expensive, exceeding the insurance LeMessurier carried.

LeMessurier faced an ethical dilemma involving a conflict between his responsibilities to ensure the safety of his building for the sake of people who use it, his responsibilities to various financial constituencies, and his self-interest, which might be served by remaining silent. What to do? He retreated to his summerhouse on an island on Sebago Lake in Maine. There, in the quiet, he worked once more through all the design and wind tunnel numbers. Suddenly he realized, with "an almost giddy sense of power," that only he could prevent an eventual disaster by taking the initiative.

Having made a decision, he acted quickly. He and Stubbins met with their insurers, lawyers, the bank management, and the city building department to describe the problem. A retrofit plan was agreed on: The wind braces would be strengthened at critical locations "by welding two-inch-thick steel plates over each of more than 200 bolted joints." Journalists, at first curious about the many lawyers converging on the various offices, disappeared when New York's major newspapers were shut down by a strike. The lawyers sought the advice of Leslie Robertson, a structural engineer with experience in disaster management. He alerted the mayor's Office of Emergency Management and the Red Cross so the surroundings of the building could be evacuated in case of a high wind alert. He also arranged for a network of strain gages to be attached to the structure at strategic points. This instrumentation allowed actual strains experienced by the steel to be monitored at a remote location. LeMessurier insisted on the installation of an emergency generator to assure uninterrupted availability of the damper. The retrofit and the tuned mass damper had been readied to withstand as much as a 200-year storm.

The parties were able to settle out of court with Stubbins held blameless; LeMessurier and his joint-venture partners were charged the $2 million his insurance agreed to pay. The total repair bill had amounted to more than $12.5 million. In acting responsibly, LeMessurier saved lives and preserved his integrity, and his professional reputation was enhanced rather than tarnished by the episode.

Meanings of *Responsibility*

If we say that LeMessurier was responsible, as a person and as an engineer, we might mean several things: He met his responsibilities (obligations); he was responsible (accountable) for doing so; he acted responsibly (conscientiously); and he is admirable (praiseworthy). Let us clarify these and related senses of *responsibility*, beginning with obligations—the core idea around which all the other senses revolve.[13]

1. *Obligations*. Responsibilities are *obligations*—types of actions that are morally mandatory. Some obligations are incumbent on each of us, such as to be honest, fair, and decent. Other obligations are *role responsibilities*, acquired when we take on special roles such as parents, employees, or professionals. Thus, a safety engineer might have responsibilities for making regular inspections at a building site, or an operations engineer might have responsibilities for identifying potential benefits and risks of one system as compared with another.

2. *Accountable*. Being responsible means being morally accountable. This entails having the general capacities for moral

---

[13] Graham Haydon, "On Being Responsible," *Philosophical Quarterly* 28 (1978): 46–57; and R. Jay Wallace, *Responsibility and the Moral Sentiments* (Cambridge, MA: Harvard University Press, 1996).

agency, including the capacity to understand and act on moral reasons. It also entails being answerable for meeting particular obligations, that is, liable to be held to account by other people in general or by specific individuals in positions of authority. We can be called on to explain why we acted as we did, perhaps providing a justification or perhaps offering reasonable excuses. We also hold ourselves accountable for meeting our obligations, sometimes responding with emotions of self-respect and pride, other times responding with guilt for harming others and shame for falling short of our ideals.

Wrongdoing takes two primary forms: voluntary wrongdoing and unintentional negligence. Voluntary wrongdoing occurs when we knew we were doing wrong and were not coerced. Sometimes it is caused by recklessness, that is, flagrant disregard of known risks and responsibilities. Other times it is a result of weakness of will, whereby we give in to temptation or fail to try hard enough. In contrast, unintentional negligence occurs when we unintentionally fail to exercise due care in meeting responsibilities. We might not have known what we were doing, but we should have known. Shoddy engineering because of sheer incompetence usually falls into this category.

3. *Conscientious, integrity.* Morally admirable engineers such as LeMessurier accept their obligations and are conscientious in meeting them. They diligently try to do the right thing, and they largely succeed in doing so, even under difficult circumstances. In this sense, being responsible is a virtue—an excellence of character. Of course, no one is perfect, and we might be conscientious in some areas of life, such as our work, and less conscientious in other areas, such as raising a child.

4. *Blameworthy/Praiseworthy.* In contexts where it is clear that accountability for wrongdoing is at issue, "responsible" becomes a synonym for *blameworthy.* In contexts where it is clear that right conduct is at issue, "responsible" is a synonym for *praiseworthy.* Thus, the question "Who is responsible for designing the antenna tower?" might be used to ask who is blameworthy for its collapse or who deserves credit for its success in withstanding a severe storm.

The preceding meanings all concerned moral responsibility, in particular as it bears on professional responsibility. Moral responsibility is distinguishable from causal, job, and legal responsibility. *Causal responsibility* consists simply in being a cause of some event. (A young child playing with matches causes a house to burn down, but the adult who left the child with the matches is morally responsible.) *Job responsibility* consists of one's assigned tasks at the place of employment. And *legal responsibility* is whatever the law requires—including legal obligations and accountability for meeting them.

The causal, job, and legal responsibilities of engineers overlap with their moral and professional responsibilities, although not completely. In particular, professional responsibilities transcend narrow job assignments. For example, LeMessurier recognized and accepted a responsibility to protect the public even though his particular job description left it unclear exactly what was required of him.

Engineering as a Profession

We have been speaking of engineering as a profession, but what exactly is a profession? In a broad sense, a profession is any occupation that provides a means by which to earn a living. In the sense intended here, however, professions are those forms of work involving advanced expertise, self-regulation, and concerted service to the public good.[14]

1. *Advanced expertise.* Professions require sophisticated skills (knowing-how) and theoretical knowledge (knowing-that) in exercising judgment that is not entirely routine or susceptible to mechanization. Preparation to engage in the work typically requires extensive formal education, including technical studies in one or more areas of systematic knowledge as well as some broader studies in the liberal arts (humanities, sciences, arts). Generally, continuing education and updating knowledge are also required.

2. *Self-regulation.* Well-established societies of professionals are allowed by the public to play a major role in setting standards for admission to the profession, drafting codes of ethics, enforcing standards of conduct, and representing the profession before the public and the government. Often this is referred to as the "autonomy of the profession," which forms the basis for individual professionals to exercise autonomous professional judgment in their work.

3. *Public good.* The occupation serves some important public good, or aspect of the public good, and it does so by making a concerted effort to maintain high ethical standards throughout the profession. For example, medicine is directed toward promoting health, law toward protecting the public's legal rights, and engineering toward technological solutions to problems concerning the public's well-being, safety, and health. The aims and guidelines in serving the public good are detailed in professional codes of ethics.

---

[14] Michael Bayles, *Professional Ethics*, 2nd ed. (Belmont, CA: Wadsworth, 1989).

Exactly how these criteria are understood and applied involves judgments about value standards. Indeed, some critics argue that the attempt to distinguish professions from other forms of work is an elitist attempt to elevate the prestige and income of certain groups of workers. Innumerable forms of work contribute to the public good, even though they do not require advanced expertise: for example, hair cutting, selling real estate, garbage collection, and professional sports. In reply, we agree that these are valuable forms of work and that professionalism should not be primarily about social status. Nevertheless, concerted efforts to maintain high standards of moral responsibility, together with a sophisticated level of required skill and the requisite autonomy to do so, warrants the recognition traditionally associated with the word *profession*.

Professions, as structured groups of professionals, have collective responsibilities to promote responsible conduct by their members. They can do so in many ways, explored throughout this book. One important way is to promulgate and take seriously a code of ethics for members of the profession, a topic we take up in Chapter 2. Taking ethics seriously involves developing procedures for disciplining irresponsible engineers, but the main emphasis in ethics should be supporting responsible individuals. In fact, the vast majority of engineers are morally committed, but they need the support of morally committed professional societies. Professions and professionals also need to think in terms of *preventive ethics*—that is, ethical reflection and action aimed at preventing moral harm and unnecessary ethical problems.

Ethical Corporations

From its inception as a profession, as distinct from a craft, much engineering has been embedded in corporations. That is because of the nature of engineering, both in its goal of producing economical and safe products for the marketplace and in its usual complexity of large projects that requires that many individuals work together. Engineer and historian Edwin T. Layton Jr. suggests that corporate control underlies the primary ethical dilemmas confronted by engineers: "The engineer's problem has centered on a conflict between professional independence and bureaucratic loyalty," and "the role of the engineer represents a patchwork of compromises between professional ideals and business demands."[15]

We will encounter ethical dilemmas that provide some support for Layton's generalization. But we add three caveats. First,

---

[15] Edwin T. Layton Jr., *The Revolt of the Engineers: Social Responsibility and the American Engineering Profession* (Baltimore: Johns Hopkins University Press, 1986), 1, 5.

corporate influence is by no means unique to engineering. Today, all professions are interwoven with corporations, including medicine, law, journalism, and science. Second, corporations make possible the goods generated by engineering, as well as giving rise to some of the ethical dilemmas they face. Third, most corporations at least strive to be morally responsible. Professional ethics and business ethics should be connected from the outset, although by no means equated.[16]

To be sure, some corporations are corrupt. Beginning in 2001, revelation of a wave of scandals shook Americans' confidence in corporations. In that year, Enron became the largest bankruptcy in U.S. history, erasing approximately $60 billion in shareholder value.[17] Created in 1985, Enron grew rapidly, selling natural gas and wholesale electricity in a new era of government deregulation. In the 1990s it began using fraudulent accounting practices, partly indulged by auditors from Arthur Andersen, a major accounting firm that collapsed in the aftermath of the Enron scandal. Enron created "Special Purpose Entities," nicknamed "Raptors" after the dinosaurs portrayed in the movie Jurassic Park—off–balance sheet partnerships designed to conceal hundreds of millions of dollars in debt and to inflate reported profits. Other unethical practices included price manipulation in sales of electricity to California resulting in massive financial losses to the state. For a time, the game of smoke and mirrors worked, keeping Enron's credit rating buoyant so that it could continue to borrow and invest heavily in ever-expanding markets, often where it lacked expertise. Indeed, for five consecutive years, between 1996 and 2000, Enron was voted in a *Fortune* magazine poll to be the most innovative corporation in the United States.

Fortunately, most corporations are not like Enron. Many, indeed most, companies place a high priority on concern for worthwhile products and ethical procedures. Since the 1960s, a "social responsibility movement" has raised attention to product quality, the well-being of workers, the wider community, and the environment. The movement is reflected in what is called "stakeholder theory": corporations have responsibilities to all groups that have a vital stake in the corporation, including employees, customers, dealers, suppliers, local communities, and the general public.[18]

---

[16] Howard Gardner, Mihaly Csikszentmihalyi, and William Damon, *Good Work: When Excellence and Ethics Meet* (New York: Basic Books, 2001).

[17] Peter C. Fusaro and Ross M. Miller, *What Went Wrong at Enron?* (New York: John Wiley & Sons, 2002); Loren Fox, *Enron: The Rise and Fall* (New York: John Wiley & Sons, 2003).

[18] R. Edward Freeman, "The Politics of Stakeholder Theory," *Business Ethics Quarterly* 4 (1994): 409–21. See also Ronald M. Green, *The Ethical Manager* (New York: Macmillan, 1994), 25–42.

The social responsibility movement in business is not without its critics who contend that corporations should concentrate solely on maximizing profits for stockholders and that there are no additional responsibilities to society, customers, and employees. In a famous essay, "The Social Responsibility of Business is to Increase Its Profits," Nobel Laureate economist Milton Friedman attacked the social responsibility movement. He argued that the paramount, indeed the sole, responsibility of management is to satisfy the desires of stockholders who entrust corporations with their money to maximize return on their investment. Management acts irresponsibly and violates stockholders' trust when it adopts further social goals, such as protecting the environment, training disadvantaged workers, using affirmative action hiring practices, or making philanthropic donations to local communities or the arts. The responsibility of managers is "to conduct the business in accordance with their [stockholders'] desires, which generally will be to make as much money as possible while conforming to the basic rules of the society, both those embodied in law and those embodied in ethical custom."[19]

Ironically, Friedman's allusion to heeding "ethical custom" invites recognition of the wider corporate responsibilities he inveighs against. In our society, the public expects corporations to contribute to the wider community good and to protect the environment, and that becomes a moral custom (as indeed it largely has). It seems clear, however, that by "ethical custom" Friedman means only refraining from fraud and deception, and he opposes everything but the most minimum regulation of business needed to protect contracts.

In its extreme form, Friedman's view is self-defeating. As quickly as the public learns that corporations are indifferent to anything but profit, it will pass restrictive laws that make profit-making difficult. Conversely, when the public perceives corporations as having wider social commitments, it is more willing to cooperate with them to assure reasonable regulations and to selectively purchase products from such socially responsible corporations. Even many investors will be more likely to stay with companies whose ethical commitments promise long-lasting success in business. For these reasons, it would be difficult to find a CEO today who would publicly say that maximum profits are all his or her company is devoted to, that greed is good, and that

---

[19] Milton Friedman, "The Social Responsibility of Business Is to Increase Its Profits," *New York Times Magazine*, September 13, 1970. See also *Capitalism and Freedom* (Chicago: University of Chicago Press, 1963), 133. For critiques of Friedman's view, see Peter Drucker, *An Introductory View of Management* (New York: Harper & Row, 1974), 61; and Robert C. Solomon, *Ethics and Excellence: Cooperation and Integrity in Business* (New York: Oxford University Press, 1992), 120.

the environment, workers, and customer safety are mere means to profit.

Sound ethics and good business go together, for the most part and in the long run. Hence at a fundamental level, the moral roles of engineers and their corporations are symbiotic, despite occasional tensions between engineers and managers. As a result of their different experience, education, and roles, higher management tends to emphasize corporate efficiency and productivity—the bottom line. Engineers and other professionals tend to emphasize excellence in creating useful, safe, and quality products. But these differences should be a matter of emphasis rather than opposition.

Senses of Corporate Responsibility

We have been talking about corporate responsibility, but the word *responsibility* is ambiguous, as noted earlier. All the senses we distinguished in connection with individuals also apply to corporations.

1. Just as individuals have responsibilities (obligations), so do corporations. To be sure, corporations are communities of individuals, structured within legal frameworks. Yet corporations have internal structures consisting of policy manuals and flowcharts assigning responsibilities to individuals.[20] When those individuals act (or should act) in accordance with their assigned responsibilities, the corporation as a unity can be said to act. Thus, when we say that Intel created a new subsidiary, we understand that individuals with the authority took certain steps.

2. Just as individuals are accountable for meeting their obligations, so corporations are accountable to the general public, to their employees and customers, and to their stockholders. Corporations, too, have the capacity for morally responsible agency because it is intelligible to speak of the corporation as acting. The actions of the corporation are performed by individuals and subgroups within the corporation, according to how the flowchart and policy manual specify areas of authority.

3. Just as individuals manifest the virtue of responsibility when they regularly meet their obligations, so too corporations manifest the virtue of responsibility when they routinely meet their obligations. In general, it makes sense to ascribe virtues such as honesty, fairness, and public spiritedness to certain corporations and not to others.

---

[20] Peter A. French, *Corporate Ethics* (New York: Harcourt Brace, 1995).

4. In contexts where it is clear that accountability for wrongdoing is at issue, "responsible" becomes a synonym for blameworthy, and in contexts where it is clear that right conduct is at issue, "responsible" is a synonym for praiseworthy. This is as true for corporations as it is for individuals.

All these moral meanings are distinct from causal responsibility, which consists simply in being a cause of some event. The meanings are also distinct from legal responsibility, which is simply what the law requires. Engineering firms can be held legally responsible for harm that was so unlikely and unforeseeable that little or no moral responsibility is involved. One famous court case involved a farmer who lost an eye when a metal chip flew off the hammer he was using.[21] He had used the hammer without problems for eleven months before the accident. It was constructed from metals satisfying all the relevant safety regulations, and no specific defect was found in it. The manufacturer was held legally responsible and required to pay damages. The basis for the ruling was the doctrine of strict legal liability, which does not require proof of defect or negligence in design. Yet surely the manufacturing firm was not morally guilty or blameworthy for the harm done. It is morally responsible only insofar as it has an obligation (based on the special relationship between it and the farmer created by the accident) to help repair, undue, or compensate for the harm caused by the defective hammer.

Discussion Questions

1. As soon as he identified the structural danger in the Citicorp building, should LeMessurier have notified the workers in the building, surrounding neighbors, and the general public who might do business in the building? Or was it enough that he made sure evacuation plans were in place and that he was prepared to provide warning to people affected in the event of a major storm?

2. Laws play an enormously important role in engineering, but sometimes they overshadow and even threaten morally responsible conduct. Thus, attorneys often advise individuals not to admit responsibility. Bring to mind some occasions where that is good advice. Then discuss whether it would have been sound advice to LeMessurier in the Citicorp Tower case.

3. Michael Davis defines professions as follows: "A profession is a number of individuals in the same occupation voluntarily organized to earn a living by openly serving a certain moral ideal

---

[21] Richard C. Vaughn, *Legal Aspects of Engineering*, 3rd ed. (Dubuque, IA: Kendall/Hunt, 1977), 41–47.

in a morally permissible way beyond what law, market, and [everyday] morality would otherwise require."[22] He argues that carpenters, barbers, porters, and other groups who organize their work around a shared code of ethics should be recognized as professionals. Do you agree or disagree, and why? Can this issue be settled by reference to a dictionary?

4. Disputes arise over how a person becomes or should become a member of an accepted profession. Such disputes often occur in engineering. Each of the following has been proposed as a criterion for being a "professional engineer" in the United States. Assess these definitions to determine which, if any, captures what you think should be part of the meaning of "engineers."

   a. Earning a bachelor's degree in engineering at a school approved by the Accreditation Board for Engineering and Technology. (If applied in retrospect, this would rule out Leonardo da Vinci, Thomas Edison, and Nikola Tesla.)

   b. Performing work commonly recognized as what engineers do. (This rules out many engineers who have become full-time managers but embraces some people who do not hold engineering degrees.)

   c. In the United States, being officially registered and licensed as a professional engineer (PE). Becoming registered typically includes: (1) passing the Engineer-in-Training Examination or Professional Engineer Associate Examination shortly before or after graduation from an engineering school, (2) working four to five years at responsible engineering, (3) passing a professional examination, and (4) paying the requisite registration fees. (Only those engineers whose work directly affects public safety and who sign official documents such as drawings for buildings are required to be registered as PEs. Engineers who practice in manufacturing or teach at engineering schools are exempt. Nevertheless, many acquire their PE licenses out of respect for the profession or for prestige.)

   d. Acting in morally responsible ways while practicing engineering. The standards for responsible conduct might be those specified in engineering codes of ethics or an even fuller set of valid standards. (This rules out scoundrels, no matter how creative they may be in the practice of engineering.)

5. Milton Friedman argues that the sole responsibility of managers is to stockholders, to maximize their profits within the bounds of law and without committing fraud. An alternative view is stake-

---

[22] Michael Davis, *Profession, Code, and Ethics* (Burlington, VT: Ashgate Publishing Company, 2002), 3.

holder theory: Managers have responsibilities to all individuals and organizations that make contracts with a corporation or otherwise are directly affected by them.[23] Clarify what you see as the implications of these alternative views as they apply to decisions about relocating a manufacturing facility to lower costs for workers' salaries. Then present and defend your view as to which of these positions is the more defensible morally.

6. Enron CEO Kenneth Lay betrayed his employees by strongly encouraging them to purchase Enron stock, even after he knew the stock was in trouble—indeed, because he knew it was in trouble—and had begun to sell large amounts of his own shares. In addition, when the stock meltdown began, a company policy prevented employees from selling their stock until it became worthless, thereby causing huge losses in employee retirement programs. Friedman and stakeholder theory would join in condemning such practices. What might each say, however, about Enron's "rank and yank" program? According to one account, every six months all employees were ranked on a 1-to-5 scale, with managers forced to place 15 percent of employees in the lowest category?[24] Those ranked lowest were given six months to improve, although usually they were given severance packages, especially because at the next six-month ranking the 15 percent rule still applied. What are the pros and cons of such employee policies for sustaining both an ethical climate and excellence?

---

[23] James J. Brummer, *Corporate Responsibility and Legitimacy* (New York: Greenwood Press, 1991), 144–64.

[24] Peter C. Fusaro and Ross M. Miller, *What Went Wrong at Enron?* (New York: John Wiley & Sons, 2002), 51. See also Loren Fox, *Enron: The Rise and Fall* (New York: John Wiley & Sons, 2003).

CHAPTER 2

Moral Reasoning and Codes of Ethics

Ethical (or moral) dilemmas are situations in which moral reasons come into conflict, or in which the applications of moral values are unclear, and it is not immediately obvious what should be done. Ethical dilemmas arise in engineering, as elsewhere, because moral values are many and varied and can make competing claims. Yet, although moral dilemmas comprise the most difficult occasions for moral reasoning, they constitute a relatively small percentage of *moral choices,* that is, decisions involving moral values. The vast majority of moral choices are clear-cut, although we sometimes fail to act responsibly because of negligence and weakness of will.

We begin by illustrating how choices involving moral values enter into routine decisions during technological development, punctuated by periodic moral dilemmas. Next we discuss some aspects (or steps) in resolving ethical dilemmas, drawing on the resources of applicable codes of ethics. Later we discuss additional roles of professional codes of ethics and comment on some possible limitations in relying solely on codes for moral guidance.

2.1 Moral Choices and Ethical Dilemmas

Designing Aluminum Cans

Henry Petroski chronicles the development of aluminum beverage cans with stay-on tab openers.[1] Aluminum cans are now ubiquitous—approximately 100 billion are produced in the United States each year. The first aluminum can was designed in 1958 by Kaiser Aluminum, in the attempt to improve on heavier and more expensive tin cans. Aluminum proved ideal as a lightweight, flexible material that allowed manufacturing of

---

[1] Henry Petroski, *Invention by Design: How Engineers Get from Thought to Thing* (Cambridge, MA: Harvard University Press, 1996), 89–103.

the bottom and sides of the can from a single sheet, leaving the top to be added after the can was filled. The trick was to make the can strong enough to keep the pressurized liquid inside, while being thin enough to be cost-effective. The can also had to fit conveniently in the hand and reliably satisfy customers' needs. Design calculations solved the problem of suitable thickness of material, but improvements came gradually in shaping of the inward-dished bottom to improve stability when the can is set down, as well as to provide some leeway for expansion of the can.

The first aluminum cans, like the tin cans before them, were opened with a separate opener, which required additional manufacturing costs to make them readily available to consumers. The need for separate openers also caused inconvenience, as Ermal Fraze discovered when, forgetting an opener while on a picnic in 1959, he had to resort to using a car bumper. Fraze, who owned Dayton Reliable Tool and Manufacturing Company and was hence familiar with metal, envisioned a design for a small lever that was attached to the can but which was removed as the can opened. The idea proved workable and was quickly embraced by manufacturers. Gradual improvements were made over subsequent years to ensure easy opening and prevention of lip and nose injuries from the jagged edges of the opening.

Within a decade an unanticipated crisis arose, however, creating an ethical dilemma. Fraze had not thought through the implications of billions of discarded pull tabs causing pollution, foot injuries, and harm to fish and infants who ingested them. The dilemma was what to do to balance usefulness to consumers with protection of the environment. A technological innovation solved the dilemma in a manner that integrated all the relevant values. In 1976 Daniel F. Cudzik invented a simple, stay-attached opener of the sort familiar today. Once again, minor design improvements came as problems were identified. Indeed, the search for improvements continues today because people with arthritic fingers or long and breakable fingernails have difficulty using the current openers. All the while, of course, the broader problem of pollution from cans themselves prompted recycling programs that now recycle more than six out of ten cans (leaving room for further improvement here as well).

Petroski recounts these developments to illustrate how engineering progresses by learning from design failures—that is, designs that cause unacceptable risks or other problems. At each stage of the design process, engineers are preoccupied with what might go wrong. The hope is to anticipate and prevent failures, drawing on knowledge about past failures. Here, however, our interest is in how moral values were embedded in the design process at all stages, in addition to surfacing in explicit ethical dilemmas concerning the environment.

If we understand moral choices broadly, as decisions involving moral values, then the development of aluminum cans can be understood as a series of routine moral choices interspersed with occasional moral dilemmas. Moral values entered implicitly into the decision-making process of engineers and their managers—decisions that probably appeared to be purely technical or purely economic. This appearance is misleading, for the technical and economic decisions had moral dimensions in four general directions: safety, environmental protection, consumer usefulness, and economic benefits.

First, human safety is obviously a moral value, rooted directly in the moral worth of human beings. Some aspects of safety seem minor—slight cuts to lips and noses from poorly designed openers and minor injuries to feet in recreation areas such as beaches. But minor injuries might cause infections, and even by themselves they have some moral significance. Again, various kinds of poisoning might occur unless all materials were tested under a range of conditions, and there are potential industrial accidents during the manufacturing process. Finally, extensive testing was needed to ensure that exploding cans, although not inherently dangerous, did not cause automobile accidents when drivers were distracted while opening cans.

A second set of moral values concern the environment. Many of these values overlap with the first set, safety. Billions of detached can openers raised the level of hazards to people walking with bare feet. Injuries to fish and other wildlife posed additional concerns. Depending on one's environmental ethic, injuries to wildlife might be understood as direct moral harms to creatures recognized as having inherent worth, or instead as indirect harms to human beings. The broader problem of environmental pollution from aluminum cans and their openers required corporate action in paying for recycled materials, community action in developing the technologies for recycling, and changes in public policy and social attitudes about recycling.

Third, some moral values are masked under terms such as *useful* and *convenient* products. We tend to think of such matters as nonmoral, especially with regard to trivial things such as sipping a carbonated beverage with a pleasing taste. But there are moral connections, however indirect or minor. After all, water is a basic need, and convenient access to pleasant-tasting liquids contributes to human well-being. However slightly, these pleasures bear on human happiness and well-being, especially when considered on the scale of mass-produced products. In addition, the aesthetic values pertaining to the shape and appearance of cans have some relevance to satisfying human desires.

Finally, the economic benefits to stakeholders in the corporation have moral implications. Money matters, and it matters morally. Jobs provide the livelihood for workers and their families that make possible the material goods that contribute to happiness—and survival. The corporation's success contributes as well to the livelihood of suppliers and retailers, as well as to stockholders.

All these values—safety, environmental protection, convenience, and money—were relevant throughout the development of aluminum cans, not merely when they explicitly entered into moral dilemmas. Hence, the case illustrates how moral values permeate engineering practice.

Steps in Resolving Ethical Dilemmas

Reasonable solutions to ethical dilemmas are clear, informed, and well-reasoned. *Clear* refers to moral clarity—clarity about which moral values are at stake and how they pertain to the situation. It also refers to conceptual clarity—precision in using the key concepts (ideas) applicable in the situation. *Informed* means knowing and appreciating the implications of morally-relevant facts. In addition, it means being aware of alternative courses of action and what they entail. *Well-reasoned* means that good judgment is exercised in integrating the relevant moral values and facts to arrive at a morally desirable solution.

These characteristics of reasonable solutions also enter as steps in resolving ethical dilemmas. By "steps" we do not mean single-file movements, but instead activities that are carried out jointly and in repeating patterns. Thus, a preliminary survey of the applicable moral values and relevant facts might be followed by conceptual clarification and additional fact gathering, which in turn evince a more nuanced understanding of the applicable values and the implications of the relevant facts. In discussing the example, we will illustrate the importance of professional codes of ethics in identifying and highlighting applicable moral reasons.

A chemical engineer working in the environmental division of a computer manufacturing firm learns that her company might be discharging unlawful amounts of lead and arsenic into the city sewer.[2] The city processes the sludge into a fertilizer used by local farmers. To ensure the safety of both the discharge and the fertilizer, the city imposes restrictive laws on the discharge of lead and arsenic. Preliminary investigations convince the engineer

---

[2] This example is a variation of the case in *Gilbane Gold,* (1989), a video made by the National Society of Professional Engineers and available at NSPE, PO Box 1020, Sewickley, Pa 15143.

that the company should implement stronger pollution controls, but her supervisor tells her the cost of doing so is prohibitive and that technically the company is in compliance with the law. She is also scheduled to appear before town officials to testify in the matter. What should she do?

1. *Moral clarity: Identify the relevant moral values.* The most basic step in confronting ethical dilemmas is to become aware of them! This means identifying the moral values and reasons applicable in the situation, and bearing them in mind as further investigations are made. These values and reasons might be obligations, rights, goods, ideals (which might be desirable but not mandatory), or other moral considerations.

   Exactly how we articulate the relevant values reflects our moral outlook. Hence, the moral frameworks discussed in Chapter 3 are relevant even in stating what the ethical dilemma is. Another resource is talking with colleagues, who can help sharpen our thinking about what is at stake in the situation. But the most useful resource in identifying ethical dilemmas in engineering are professional codes of ethics, as interpreted in light of one's ongoing professional experience.

   Like most codes of ethics, the code of ethics of the American Institute of Chemical Engineers (AIChE) indicates the engineer has at least three responsibilities in the situation. One responsibility is to be honest: "Issue statements or present information only in an objective and truthful manner." A second responsibility is to the employer: "Act in professional matters for each employer or client as faithful agents or trustees, avoiding conflicts of interest and never breaching confidentiality." A third responsibility is to the public, and also to protect the environment: "Hold paramount the safety, health, and welfare of the public and protect the environment in performance of their professional duties." In the case at hand, the members of the public most directly affected are the local farmers, but the dangerous chemicals could affect more persons as lead and arsenic are drawn into the food chain. Additional moral considerations, not cited in the code, include duties to maintain personal and professional integrity, and rights to pursue one's career.

2. *Conceptual clarity: Be clear about key concepts.* Professionalism requires being a faithful agent of one's employer, but does that mean doing what one's supervisor directs or doing what is good for the corporation in the long run? These might be different things, in particular when one's supervisor is adopting a short-term view that could harm the long-term interests of the corporation. Again, what does it mean to "hold paramount the safety, health, and welfare of the public" in the case

at hand? Does it pertain to all threats to public health, or just serious threats, and what is a "serious" threat? Again, does being "objective and truthful" simply mean never lying (intentionally stating a falsehood), or does it mean revealing all pertinent facts (withholding nothing important) and doing so in a way that gives no preference to the interests of one's employer over the needs of the public to be informed of hazards?

3. *Informed about the facts: Obtain relevant information.* This means gathering information that is pertinent in light of the applicable moral values (as identified in step 1). Sometimes the primary difficulty in resolving moral dilemmas is uncertainty about the facts, rather than conflicting values per se. Certainly in the case at hand, the chemical engineer needs to check and recheck her findings, perhaps asking colleagues for their perspectives. Her corporation seems to be violating the law, but is it actually doing so? We, like the engineer, need to know more about the possible harm caused by the minute quantities of lead and arsenic over time. How serious is it, and how likely to cause harm?

4. *Informed about the options: Consider all (realistic) options.* Initially, ethical dilemmas seem to force us into a two-way choice: Do this or do that. Either bow to a supervisor's orders or blow the whistle to the town authorities. A closer look often reveals additional options. (Sometimes writing down the main options and suboptions as a matrix or decision tree ensures that all options are considered.) The chemical engineer might be able to suggest a new course of research that will improve the removal of lead and arsenic. Or she might discover that the city's laws are needlessly restrictive and should be revised. Perhaps she can think of a way to convince her supervisor to be more open-minded about the situation, especially given the possible damage to the corporation's image if it should later be found in violation of the law. Unless an emergency develops, these and other steps should be attempted before informing authorities outside the corporation—a desperate last resort, especially given the likely penalties for whistle-blowing (see Chapter 7).

5. *Well-reasoned: Make a reasonable decision.* Arrive at a carefully reasoned judgment by weighing all the relevant moral reasons and facts. This is not a mechanical process that a computer or algorithm might do for us. Instead, it is a deliberation aimed at integrating all the relevant reasons, facts, and values—in a morally reasonable manner. If there is no ideal solution, as is often the case, we seek a satisfactory one, what Herbert Simon dubbed "satisficing."

Often a code of ethics provides a straightforward solution to dilemmas, but not always. Codes are not recipe books that contain a comprehensive list of absolute (exceptionless) rules together with precise hierarchies of relative stringency among the rules. What about the case at hand? The code does assert one very important hierarchy: *Hold paramount the public safety, health, and welfare*. The AIChE code also requires engineers to "formally advise their employers or clients (and consider further disclosure, if warranted) if they perceive that a consequence of their duties will adversely affect the present or future health or safety of their colleagues or the public." This statement, combined with the statement of the paramount responsibility, makes it clear that the responsibility to be a faithful agent of the employer does not override professional judgment in important matters of public safety.

At the same time, the recommendation to "consider further disclosure, if warranted" / seems somewhat lukewarm, both because it is placed parenthetically and because it only says "consider." It suggests something to think about, rather than a firm statement of duty. As such, it is weaker than statements in some other codes, including the code of the National Society of Professional Engineers (NSPE), that require notification of appropriate authorities when one's judgment is overridden in matters where public safety is endangered. Which of these codes takes precedence?

Furthermore, exactly what does the paramount statement entail in the case at hand? If the engineer is convinced her company produces valuable computers, might she reasonably conclude that the public good is held paramount by coming "close enough" to obeying the law? As for the requirement to be "objective and truthful," that certainly implies not lying to the town officials, but might she reasonably conclude she is being objective by not divulging information her supervisor says is confidential? Obviously, such conclusions might be products of rationalization (biased reasoning), rather than sound moral reasoning. We mention them only to suggest that codes are no substitute for morally good judgment—honest, fair, responsible moral judgment. Indeed, as we have just seen, good judgment is needed even in interpreting the code of ethics.[3] The development of good moral judgment is part and parcel of developing experience in engineering. It is also a primary goal in studying ethics.

---

[3] On interpreting codes intelligently, see Michael Davis, "Professional Responsibility as Just Following the Rules," in Michael Davis *Profession, Code, and Ethics* (Burlington, VT: Ashgate Publishing Company, 2002), 83–98.

Right-Wrong or Better-Worse?

We might divide ethical dilemmas into two broad categories. On the one hand, many dilemmas have solutions that are either right or wrong. "Right" means that one course of action is obligatory, and failing to do that action is unethical (immoral). In most instances a code of ethics specifies what is clearly required: Obey the law and heed engineering standards, do not offer or accept bribes, speak and write truthfully, maintain confidentiality, and so forth. On the other hand, some dilemmas have two or more reasonable solutions, no one of which is mandatory, but one of which should be chosen. These solutions might be better or worse than others in some respects but not necessarily in all respects.

In illustrating the two types of dilemmas, we will continue discussing the requirement to hold paramount the safety, health, and welfare of the public. We will also draw on examples from the NSPE Board of Ethical Review (BER). This board provides the valuable service of applying the NSPE code to cases that are fictionalized but based on actual events. The board's rulings are published periodically in bound volumes, and they are also available on the Internet (http://www.niee.org). Although the cases tend to emphasize consulting rather than corporate engineering, they provide illuminating examples about how to intelligently interpret the NSPE code. They are intended solely for educational purposes, to stimulate reflection and discussion. Consider BER Case 93–7:

> Engineer A, an environmental engineer, is retained by a major industrial owner to examine certain lands adjacent to an abandoned industrial facility formerly owned and operated by the owner. Owner's attorney, Attorney X, requests that as a condition of the retention agreement that Engineer A sign a secrecy provision whereby Engineer A would agree not to disclose any data, findings, conclusions, or other information relating to his examination of the owner's land to any other party unless ordered by a court. Engineer A signs the secrecy provision.[4]

What is the ethical problem? Although the NSPE code does not explicitly forbid signing the secrecy provision, it does in fact require engineers to hold paramount the public safety and, if their judgment should be overruled in matters of public safety, to notify proper authorities. This implies that Engineer A should not sign

---

[4] National Society of Professional Engineers, *Opinions of the Board of Ethical Review*, vol. VII (Alexandria, VA: National Society of Professional Engineers, 1994), Case No. 93–7, 101.

a secrecy provision that precludes acting according to the code. As the BER states, "We do not believe an engineer should ever agree, either by contract or other means, to relinquish his right to exercise professional judgment in such matters." The board also cites the provisions in the code requiring confidentiality about clients, not only proprietary (legally protected) information, but all information obtained in the course of providing professional services. Nevertheless, the paramount clause requires that the public safety, health, and welfare be an overriding consideration. The spirit, if not the letter, of the code indicates that it is unethical for Engineer A to sign the secrecy provision.

As it stands, the decision about whether to sign the secrecy agreement was a dilemma involving lack of clarity about how two moral values applied in the situation: confidentiality and the paramount responsibility to protect the public safety, health, and welfare. (Similar dilemmas arise concerning restrictive confidentiality agreements between salaried engineers and their corporations, although engineers and their corporations are usually granted much wider leeway in reaching confidentiality agreements.) According to NSPE, the solution to this dilemma involves one mandatory action: Refrain from signing the agreement.

But Engineer A does sign the secrecy agreement, and so what happens at that point? The board does not address itself to this question, but clearly another ethical dilemma arises: A commitment and perhaps an obligation to keep the agreement is created, but the paramount responsibility still applies. Hence, if dangers to the public are discovered and if the client refuses to remedy them, the engineer would be obligated to notify proper authorities. But should Engineer A go back to the client and ask to have the secrecy provision revoked? And if the client refuses, should Engineer A break the contract, a step that might have legal repercussions? Or should Engineer A simply hope that no problems will arise and continue with his or her contracted work, postponing any hard decisions until later? As these questions indicate, dilemmas can generate further dilemmas! In this instance, possibly more than one option is reasonable—if not ideal, at least permissible.

To underscore the possibility of several solutions, no one of which is ideal in every regard, consider another case, BER Case 96–4.

Engineer A is employed by a software company and is involved in the design of specialized software in connection with the operations of facilities affecting the public health and safety (i.e., nuclear, air quality control, water quality control). As part of the design of a particular software system, Engineer A conducts extensive testing, and although the tests demonstrate that the software is safe to use under existing standards, Engineer A is aware of new draft standards that are about to be released by a standard setting organization—standards which the newly designed software may not meet. Testing is extremely costly and the company's clients are eager to begin to move forward. The software company is eager to satisfy its clients, protect the software company's finances, and protect existing jobs; but at the same time, the management of the software company wants to be sure that the software is safe to use. A series of tests proposed by Engineer A will likely result in a decision whether to move forward with the use of the software. The tests are costly and will delay the use of the software at least six months, which will put the company at a competitive disadvantage and cost the company a significant amount of money. Also, delaying implementation will mean the state public service commission utility rates will rise significantly during this time. The company requests Engineer A's recommendation concerning the need for additional software testing.[5]

Here the answer seems obvious enough. In tune with our conviction that good engineering and ethics go together, Engineer A should write an honest report. Indeed, it might seem that there is no dilemma for Engineer A at all because what should be done is so obvious. To be sure, the software company faces an ethical dilemma: Is it all right to proceed without the additional testing? But that is a dilemma for the managers, it would seem, not the engineer. The engineer should focus solely on safety issues and fully inform management about the risks, the new draft standards, and the proposed tests. That is what the BER concludes: "Engineer A has a professional obligation under the Code of Ethics to explain why additional testing is required and to recommend to his company that it be undertaken. By so doing, the company can make an informed decision about the need for additional testing and its effects on the public health, safety, and welfare."

In reaching this conclusion, the board suggests the engineer should focus solely on safety, leaving consideration of other nontechnical matters (such as financial impacts) to management. Yet the board also concludes that the recommendation should be for further testing. As authors, we do not find that conclusion altogether obvious from the facts presented. Much depends on exactly what the risks and circumstances are, and here we need further information. In our view, the case illustrates how there can be better or worse decisions, both of which might be permissible in the situation. Moreover, one decision might be better in some respects, and the other decision better in other respects.

---

[5] National Society of Professional Engineers, *Opinions of the Board of Ethical Review*, Case 96–4, http://www.niee.org/cases/case96–4.

Perhaps the public health and safety might well be served by having the company do the further tests even at the risk of severe economic hardship or even bankruptcy. It would be better, however, for employees and customers that this not occur. The paramountcy clause apparently requires bankruptcy rather than imposing unacceptable and severe risks on the public, but it is unclear that such risks are posed in this case. Hence, there might be two morally permissible courses of action: Do the tests; do not do the tests. Each option might have further options under it. For example, do the tests, but interrupt them if economic conditions worsen; or do the tests, but devise a quicker version of them; or do the tests, but go ahead with the present sale, being willing to make modifications if the tests raise concerns.

Moral Decision Making as Design

We have been discussing engineering design as a domain where moral choices are made. Turning things around, some thinkers suggest that engineering design provides an illuminating model for thinking about all moral decision making, not just decisions within engineering.

Thus, John Dewey (1859–1952) used engineering as a metaphor for thinking about moral reasoning in general.[6] Dewey was an exponent of pragmatism, a distinctively American ethical outlook that emphasized intelligent choices made in light of (1) attention to the practical contexts in which moral issues arise and find their solutions, (2) imaginative awareness of wider contexts that illuminate dilemmas, (3) reasonably integrating multiple claims, and (4) experimenting to find an optimal solution. He referred to his pragmatic theory of inquiry as "instrumentalism," but late in life he said that "technological" is a better term for what he had in mind.[7]

More recently and more fully, Caroline Whitbeck suggests that engineering design is in many respects a model for "designing" courses of action in many moral situations, in engineering and elsewhere.[8] As an illustration, she cites a class assignment in which she supervised several mechanical engineering students. The assignment was to design a child seat that fits on top of standard suitcases with wheels. She specified several constraints. Some pertained to size: The child seat must be easily removable and storable under airplane seats and in overhead storage

---

[6] John Dewey, *Human Nature and Conduct* (New York: Modern Library, 1957). See also James D. Wallace, *Moral Relevance and Moral Conflict* (Ithaca, NY: Cornell University Press, 1988).

[7] Cf. Larry A. Hickman, *John Dewey's Pragmatic Technology* (Bloomington, IN: Indiana University Press, 1990), 58.

[8] Caroline Whitbeck, *Ethics in Engineering Practice and Research* (New York: Cambridge University Press, 1998), 53–68.

bins. Others pertained to use: The seat must have multiple uses, including the possibility of strapping it into a seat on an airplane. Still others set safety limits: conformity to applicable safety laws plus avoiding unnecessary dangers. Yet there were many areas of uncertainty and ambiguity surrounding how to maximize safety (for example, when carrying the infant in the seat) and how many convenience features to include, such as storage spaces for baby bottles and diapers.

The students arrived at strikingly different designs, varying in size and shape as well as in the basic structure of the crossbar that held the infant in place. Several were reasonable solutions to the design problem. Yet no design was ideal in every regard, and each had strengths and weaknesses. For example, one was larger and would accommodate older infants, but the added size increased the cost of manufacturing. Again, the bar securing the infant was more convenient in some directions of motion and less convenient in other directions.

Whitbeck identifies five aspects of engineering decisions that highlight important aspects of many moral decisions in general. First, usually there are alternative solutions to design problems, more than one of which is satisfactory or "satisfices."

Second, multiple moral factors are involved, and among the satisfactory solutions for design problems, one solution is typically better in some respects and less satisfactory in other respects when compared with alternative solutions.

Third, some design solutions are clearly unacceptable. Designs of the child seat that violate the applicable laws or impose unnecessary hazards on infants are ruled out. In general, there are many "background constraints," for example justice and decency, which limit the range of reasonable moral options.

Fourth, engineering design often involves uncertainties and ambiguities, not only about what is possible and how to achieve it, but also about the specific problems that will arise as solutions are developed.

Finally, design problems are dynamic. In the real world the design of the child seat would go through much iteration, as feedback was received from testing and use of the child seat.

Discussion Questions

With regard to each of the following cases, answer several questions. First, what is the moral dilemma (or dilemmas), if any? In stating the dilemma, make explicit the competing moral reasons involved. Second, are there any concepts (ideas) involved in dealing with the moral issues that it would be useful to clarify? Third, what factual inquiries do you think might be needed in making a reliable judgment about the case? Fourth, what are the options you see available for solving the dilemma? Fifth, which of these options is required (obligatory, all things considered) or permissible (all right)?

Case 1. An inspector discovers faulty construction equipment and applies a violation tag, preventing its continued use. The inspector's supervisor, a construction manager, views the case as a minor infraction of safety regulations and orders the tag removed so the project will not be delayed. What should she do?

Case 2. A software engineer discovers that a colleague has been downloading restricted files that contain trade secrets about a new product that the colleague is not personally involved with. He knows the colleague has been having financial problems, and he fears the colleague is planning to sell the secrets or perhaps leave the company and use them in starting up his own company. Company policy requires him to inform his supervisor, but the colleague is a close friend. Should he first talk with the friend about what he is doing, or should he immediately inform his supervisor?

Case 3. An aerospace engineer is volunteering as a mentor for a high school team competing in a national contest to build a robot that straightens boxes. The plan was to help the students on weekends for at most eight to ten hours. As the national competition nears, the robot's motor overheats, and the engine burns out. He wants to help the dispirited students and believes his mentoring commitment requires he do more. But doing so would involve additional evening work that could potentially harm his work, if not his family.

Case 4. During an investigation of a bridge collapse, Engineer A investigates another similar bridge, and finds it to be only marginally safe. He contacts the governmental agency responsible for the bridge and informs them of his concern for the safety of the structure. He is told that the agency is aware of this situation and has planned to provide in next year's budget for its repair. Until then, the bridge must remain open to traffic. Without this bridge, emergency vehicles such as police and fire apparatus would have to use an alternate route that would increase their response time by approximately twenty minutes. Engineer A is thanked for his concern and asked to say nothing about the condition of the bridge. The agency is confident that the bridge will be safe.[9]

Case 5. A cafeteria in an office building has comfortable tables and chairs, indeed too comfortable: They invite people to linger

---

[9] Unpublished case study written by and used with the permission of L. R. Smith and Sheri Smith.

longer than the management desires.[10] You are asked to design uncomfortable ones to discourage such lingering.

2.2 Codes of Ethics

Importance of Codes

Codes of ethics state the moral responsibilities of engineers as seen by the profession and as represented by a professional society. Because they express the profession's collective commitment to ethics, codes are enormously important, not only in stressing engineers' responsibilities but also in supporting the freedom needed to meet them.

Codes of ethics play at least eight essential roles: serving and protecting the public, providing guidance, offering inspiration, establishing shared standards, supporting responsible professionals, contributing to education, deterring wrongdoing, and strengthening a profession's image.

1. *Serving and protecting the public.* Engineering involves advanced expertise that professionals have and the public lacks, and also considerable dangers to a vulnerable public. Accordingly, professionals stand in a fiduciary relationship with the public: Trust and trustworthiness are essential. A code of ethics functions as a commitment by the profession as a whole that engineers will serve the public health, safety, and welfare. In one way or another, the remaining functions of codes all contribute to this primary function.

2. *Guidance.* Codes provide helpful guidance by articulating the main obligations of engineers. Because codes should be brief to be effective, they offer mostly general guidance. Nonetheless, when well written, they identify primary responsibilities. More specific directions may be given in supplementary statements or guidelines, which tell how to apply the code.

3. *Inspiration.* Because codes express a profession's collective commitment to ethics, they provide a positive stimulus (motivation) for ethical conduct. In a powerful way, they voice what it means to be a member of a profession committed to responsible conduct in promoting the safety, health, and welfare of the public. Although this paramount ideal is somewhat vague, it expresses a collective commitment to the public good that inspires individuals to have similar aspirations.

4. *Shared standards.* The diversity of moral viewpoints among individual engineers makes it essential that professions establish explicit standards, in particular minimum (but hopefully high)

---

[10] The case is a variation on one described by Donald A. Norman, *The Design of Everyday Things* (New York: Doubleday, 1988), 154. Also published as *The Psychology of Everyday Things* (New York: Basic Books, 1988).

standards. In this way, the public is assured of a standard of excellence on which it can depend, and professionals are provided a fair playing field in competing for clients.

5. *Support for responsible professionals.* Codes give positive support to professionals seeking to act ethically. A publicly proclaimed code allows an engineer, under pressure to act unethically, to say: "I am bound by the code of ethics of my profession, which states that . . ." This by itself gives engineers some group backing in taking stands on moral issues. Moreover, codes can potentially serve as legal support for engineers criticized for living up to work-related professional obligations.

6. *Education and mutual understanding.* Codes can be used by professional societies and in the classroom to prompt discussion and reflection on moral issues. Widely circulated and officially approved by professional societies, codes encourage a shared understanding among professionals, the public, and government organizations about the moral responsibilities of engineers. A case in point is NSPE's BER, which actively promotes moral discussion by applying the NSPE code to cases for educational purposes.

7. *Deterrence and discipline.* Codes can also serve as the formal basis for investigating unethical conduct. Where such investigation is possible, a deterrent for immoral behavior is thereby provided. Such an investigation generally requires paralegal proceedings designed to get at the truth about a given charge without violating the personal rights of those being investigated. Unlike the American Bar Association and some other professional groups, engineering societies cannot by themselves revoke the right to practice engineering in the United States. Yet some professional societies do suspend or expel members whose professional conduct has been proven unethical, and this alone can be a powerful sanction when combined with the loss of respect from colleagues and the local community that such action is bound to produce.

8. *Contributing to the profession's image.* Codes can present a positive image to the public of an ethically committed profession. Where warranted, the image can help engineers more effectively serve the public. It can also win greater powers of self-regulation for the profession itself, while lessening the demand for more government regulation. The reputation of a profession, like the reputation of an individual professional or a corporation, is essential in sustaining the trust of the public.

Abuse of Codes

When codes are not taken seriously within a profession, they amount to a kind of window dressing that ultimately increases public cynicism about the profession. Worse, codes occasionally stifle dissent within the profession and are abused in other ways.

Probably the worst abuse of engineering codes is to restrict honest moral effort on the part of individual engineers to preserve the profession's public image and protect the status quo. Preoccupation with keeping a shiny public image may silence healthy dialogue and criticism. And an excessive interest in protecting the status quo may lead to a distrust of the engineering profession on the part of both government and the public. The best way to increase trust is by encouraging and helping engineers to speak freely and responsibly about public safety and well-being. This includes a tolerance for criticisms of the codes themselves, rather than allowing codes to become sacred documents that have to be accepted uncritically.

On rare occasions, abuses have discouraged moral conduct and caused serious harm to those seeking to serve the public. For example, two engineers were expelled from American Society of Civil Engineers (ASCE) for violating a section of its code forbidding public remarks critical of other engineers. Yet the actions of those engineers were essential in uncovering a major bribery scandal related to the construction of a dam for Los Angeles County.[11]

Moreover, codes have sometimes placed unwarranted "restraints of commerce" on business dealings to benefit those within the profession. Obviously there is disagreement about which, if any, entries function in these ways. Consider the following entry in the pre-1979 versions of the NSPE code: The engineer "shall not solicit or submit engineering proposals on the basis of competitive bidding." This prohibition was felt by the NSPE to best protect the public safety by discouraging cheap engineering proposals that might slight safety costs to win a contract. The Supreme Court ruled, however, that it mostly served the self-interest of established engineering firms and actually hurt the public by preventing the lower prices that might result from greater competition (*National Society of Professional Engineers v. United States* [1978]).

Limitations of Codes

Codes are no substitute for individual responsibility in grappling with concrete dilemmas. Most codes are restricted to gen-

---

[11] Edwin T. Layton, "Engineering Ethics and the Public Interest: A Historical View," in *Ethical Problems in Engineering*, vol. 1, ed. Albert Flores (Troy, NY: Rensselaer Polytechnic Institute, 1980), 26–29.

eral wording, and hence inevitably contain substantial areas of vagueness. Thus, they may not be able to straightforwardly address all situations. At the same time, vague wording may be the only way new technical developments and shifting social and organizational structures can be accommodated.

Other uncertainties can arise when different entries in codes come into conflict with each other. Usually codes provide little guidance as to which entry should have priority in those cases. For example, as we have noted, tensions arise between stated responsibilities to employers and to the wider public. Again, duties to speak honestly—not just to avoid deception, but also to reveal morally relevant truths—are sometimes in tension with duties to maintain confidentiality.

A further limitation of codes results from their proliferation. Andrew Oldenquist (a philosopher) and Edward Slowter (an engineer and former NSPE president) point out how the existence of separate codes for different professional engineering societies can give members the feeling that ethical conduct is more relative and variable than it actually is.[12] But Oldenquist and Slowter have also demonstrated the substantial agreement to be found among the various engineering codes, and they call for the adoption of a unified code.

Most important, despite their authority in guiding professional conduct, codes are not always the complete and final word.[13] Codes can be flawed, both by omission and commission. Until recently, for example, most codes omitted explicit mention of responsibilities concerning the environment. We also note that codes invariably emphasize responsibilities but say nothing about the rights of professionals (or employees) to pursue their endeavors responsibly. An example of commission is the former ban in engineering codes on competitive bidding. Codes, after all, represent a compromise between differing judgments, sometimes developed amidst heated committee disagreements. As such, they have a great "signpost" value in suggesting paths through what can be a bewildering terrain of moral decisions. But they

---

[12] Andrew G. Oldenquist and Edward E. Slowter, "Proposed: A Single Code of Ethics for All Engineers," *Professional Engineer* 49 (May 1979): 8–11.

[13] John Ladd, "The Quest for a Code of Professional Ethics," in Rosemary Chalk, Mark Frankel, and Sallie B. Chafer, eds., *AAAS Professional Ethics Project* (Washington, DC: American Association for the Advancement of Science, 1980), 154–59. See also Heinz C. Luegenbiehl, "Codes of Ethics and the Moral Education of Engineers," *Business and Professional Ethics Journal* 2, no. 4 (1983): 41–61. Both are reprinted in Deborah G. Johnson, ed., *Ethical Issues in Engineering* (Englewood Cliffs, NJ: Prentice Hall, 1991), 130–36 and 137–54, respectively.

should never be treated as sacred canon in silencing healthy moral debate, including debate about how to improve them.

This last limitation of codes connects with a wider issue about whether professional groups or entire societies can create sets of standards for themselves that are both morally authoritative and not open to criticism, or whether group standards are always open to moral scrutiny in light of wider values familiar in everyday life. This is the issue of ethical relativism.

Ethical Relativism

Does a profession's code of ethics create the obligations that are incumbent on members of the profession, so that engineers' obligations are entirely relative to their code of ethics? Or does the code simply record the obligations that already exist?

One view is that codes try to put into words obligations that already exist, whether or not the code is written. As Stephen Unger writes, codes "recognize" obligations that already exist: "A code of professional ethics may be thought of as a collective recognition of the responsibilities of the individual practitioners"; codes cannot be "used in cookbook fashion to resolve complex problems," but instead they are "valuable in outlining the factors to be considered."[14] Unger takes codes very seriously as a profession's shared voice in articulating the responsibilities of its practitioners. A good code provides valuable focus and direction, but it does not generate obligations so much as articulate obligations that already exist.

Michael Davis disagrees, and he places far greater emphasis on professional codes of ethics. In his view, codes are conventions established within professions to promote the public good. As such, they are morally authoritative. The code itself generates obligations: "a code of ethics is, as such, not merely good advice or a statement of aspiration. It is a standard of conduct which, if generally realized in the practice of a profession, imposes a *moral* obligation on each member of the profession to act accordingly."[15] Notice the word "imposes," as distinct from "recognizing" an obligation that already exists. To violate the code is wrong because it creates an unfair advantage in competing with other professionals in the marketplace.

Davis has been accused of endorsing *ethical relativism,* also called ethical conventionalism, which says that moral values are entirely relative to and reducible to customs—to the conventions,

---

[14] Stephen H. Unger, *Controlling Technology*, 2nd ed. (New York: John Wiley & Sons, 1994), 106.

[15] Michael Davis, *Thinking Like an Engineer* (New York: Oxford University Press, 1998), 111.

laws, and norms of the group to which one belongs.[16] What is right is simply what conforms to custom, and it is right solely *because* it conforms to customs. We can never say an act is objectively right or obligatory without qualification, but only that it is right for members of a given group because it is required by their customs. In the words of anthropologist Ruth Benedict, "We recognize that morality differs in every society, and is a convenient term for socially approved habits. Mankind has always preferred to say, 'It is morally good,' rather than 'It is habitual.' . . . But historically the two phrases are synonymous."[17] In particular, professional ethics is simply the set of conventions embraced by members of a profession, as expressed in their code.

There are problems with ethical relativism, whether we are talking about the conventions of a profession such as engineering or the conventions of a society in its entirety. By viewing customs as self-certifying, ethical relativism rules out the possibility of critiquing the customs from a wider moral framework. For example, it leaves us without a basis for criticizing genocide, the oppression of women and minorities, child abuse, torture, and reckless disregard of the environment, when these things are the customs of another culture. Regarding professional ethics, ethical relativism implies that we cannot morally critique a given code of ethics, giving reasons for why it is justified in certain ways and perhaps open to improvement in other ways.

Ethical relativism also seems to allow any group of individuals to form its own society with its own conventions, perhaps ones that common sense tells us are immoral. Again, an engineer might be a member of one or more professional societies, a weapons development corporation and a pacifist religious tradition, and the customs of these groups in matters of military work might point in different directions.

Although ethical relativism is a dubious moral outlook, it remains true that moral judgments are made "in relation to" particular circumstances, such as those of engineering. It is also true that customs are "morally relevant" (though not always decisive) in deciding how we ought to behave. Finally, *some* moral requirements are indeed established by mutual agreements. Just as laws establish the legal and moral permissibility of driving on the right side of the road (in the United States) or the left side (in

---

[16] In replying to this criticism, Davis confusingly switches from his defense of actual codes to ideal codes: "When an immoral provision appears in an actual code, it is, strictly speaking, not part of it (not, that is, what 'Obey your profession's code' commands obedience to)." Michael Davis, *Profession, Code, and Ethics,* 32.

[17] Ruth Benedict, "Anthropology and the Abnormal," *Journal of General Psychology* 10 (1934): 59–82. Also see Benedict's *Patterns of Culture* (Boston: Houghton Mifflin, 1934).

England), some requirements in engineering codes of ethics create obligations. For example, some of the specific conflicts of interest forbidden in codes of ethics are forbidden by agreement within the profession to ensure fair competition among engineers.

In our view, then, Unger and Davis are both partly correct. Unger is correct in holding that many of the entries in codes of ethics state responsibilities that would exist regardless of the code—for example, to protect the safety, health, and welfare of the public. Davis is correct that some parts of codes are conventions arrived at by mutual agreement within the profession.

Justification of Codes

If codes of ethics do not merely state conventions, as ethical relativists hold, what does justify those responsibilities that are not mere creations of convention? A code, we might say, specifies the (officially endorsed) "customs" of the professional "society" that writes and promulgates it as incumbent on all members of a profession (or at least members of a professional society). When these values are specified as responsibilities, they constitute *role responsibilities*—that is, obligations connected with a particular social role as a professional. These responsibilities are not self-certifying, any more than other customs are.

A sound professional code will stand up to three tests: (1) It will be clear and coherent; (2) it will organize basic moral values applicable to the profession in a systematic and comprehensive way, highlighting what is most important; and (3) it will provide helpful and reasonable guidance that is compatible with our most carefully considered moral convictions (judgments, intuitions) about concrete situations. In addition, it will be widely accepted within the profession.

But how can we determine whether the code meets these criteria? One way is to test the code against ethical theories of the sort discussed in Chapter 3—theories that attempt to articulate wider moral principles. Obviously, testing the code in light of an ethical theory will need to take close account of both the morally relevant features of engineering and the kinds of public goods engineering seeks to provide for the community. A codified professional ethics develops certain parts of ordinary ethics to promote the profession's public good within particular social settings. In doing so, some elements of ordinary morality take on increased importance in professional settings, as they promote the public goods served by a profession.[18] As a result, a justified professional code

---

[18] Paul F. Camenisch, *Grounding Professional Ethics in a Pluralistic Society* (New York: Haven Publications, 1983).

will take account of both the profession's public good and social frameworks and institutional settings. As these factors change, and as a profession advances, codes of ethics are revised—codes are not set in concrete.

To conclude, any set of conventions, whether codes of ethics or actual conduct, should be open to scrutiny in light of wider values. At the same time, professional codes should be taken very seriously. They express the good judgment of many morally concerned individuals, the collective wisdom of a profession at a given time. Certainly codes are a proper starting place for an inquiry into professional ethics; they establish a framework for dialogue about moral issues; and more often than not, they cast powerful light on the dilemmas confronting engineers.

Discussion Questions

1. From the Web site of an engineering professional society, select a code of ethics of interest to you, given your career plans; for example, the American Society of Civil Engineers, the American Institute of Chemical Engineers, the American Society of Mechanical Engineers, or the Institute of Electrical and Electronics Engineers. Compare and contrast the code with the NSPE code (see Appendix), selecting three or four specific points to discuss. Do they state the same requirements with the same emphasis?

2. With regard to the same two codes you used in question 1, list three examples of responsibilities that you believe would be incumbent on engineers even if the written code did not exist, and explain why. Also list two examples, if any, of responsibilities created (entirely or in part) because the code was written as a consensus document within the profession.

3. Is the following argument for ethical relativism a good argument? That is, is its premise true and does the premise provide good reason for believing the conclusion?

   a. People's beliefs and attitudes in moral matters differ considerably from society to society. (Call this statement "descriptive relativism," because it simply describes the way the world is.)

   b. Therefore, the dominant conventional beliefs and attitudes in the society are morally justified and binding (ethical relativism).

4. Reflection on the Holocaust led many anthropologists and other social scientists to reconsider ethical relativism. The Holocaust also reminds us of the power of custom, law, and social authority to shape conduct. Nazi Germany relied on the expertise of engineers, as well as other professionals, in carrying out genocide, as well as its war efforts.

   a. Do you agree that the Holocaust is a clear instance of where a cross-cultural judgment about moral wrong and right can be made?

   b. Judging actions to be immoral is one thing; blaming persons for wrongdoing is another (where blame is a morally negative attitude toward a person). Present and defend your view about whether the Nazi engineers and other professionals are blameworthy. Is blaming pointless, because the past is past? Or is cross-cultural blame, at least in this extreme instance, an important way of asserting values that we cherish?

5. Moral skeptics challenge whether sound moral reasoning is possible. An extreme form of moral skepticism is called *ethical subjectivism:* Moral judgments merely express feelings and attitudes, not beliefs that can be justified or unjustified by appeal to moral reasons. The most famous version of ethical subjectivism is called emotivism: Moral statements are merely used to express emotions—to emote—and to try to influence other people's behavior, but they are not supportable by valid moral reasons.[19] What would ethical relativists say about ethical subjectivism? What should be said in reply to the ethical subjectivist?

   Using an example, such as moral reasoning in designing aluminum cans (Petroski) or in designing a portable seat for infants (Whitbeck), discuss how moral reasons can be objective (justified) even though they sometimes allow room for different applications to particular situations.

---

[19] Charles Stevenson, *Ethics and Language* (New Haven, CT: Yale University Press, 1944).

CHAPTER 3

Moral Frameworks

An ethical theory seeks to provide a comprehensive perspective on morality that clarifies, organizes, and guides moral reflection. If successful, it provides a framework for making reasonable moral choices and resolving moral dilemmas—not a simple formula, but rather a unifying way to identify and integrate moral reasons. As one of their applications, ethical theories ground the requirements in engineering codes of ethics by reference to broader moral principles. In doing so, they illuminate connections between engineering ethics and everyday morality, that is, the justified moral values that play a role in all areas of life.

We discuss five types of ethical theories (and traditions) that have been especially influential: rights ethics, duty ethics, utilitarianism, virtue ethics, and self-realization ethics. *Rights ethics* says we ought to respect human rights, and *duty ethics* says we ought to respect individuals' rational autonomy. *Utilitarianism* says that we ought to maximize the overall good, taking into equal account all those affected by our actions. *Virtue ethics* says that good character is central to morality. *Self-realization ethics* emphasizes the moral significance of self-fulfillment. None of these theories has won a consensus, although each has proven attractive to many people. At least in some of their versions, they widely agree in their practical implications.

Rights Ethics

3.1 Rights Ethics, Duty Ethics, Utilitarianism

Rights are moral entitlements and valid moral claims that impose duties on other people. All ethical theories leave some room for rights, but the ethical theory called *rights ethics* is distinctive in that it makes human rights the ultimate appeal—the moral bottom line. Human rights constitute a moral authority to make legitimate moral demands on others to respect our choices, recognizing that others can make similar claims on us. At its core, rights ethics emphasizes respecting the inherent dignity and worth of individuals as they exercise their liberty.

Rights ethics is the most familiar ethical theory, for it provides the moral foundation of the political and legal system of the United States. Thus, in the *Declaration of Independence* Thomas Jefferson wrote: "We hold these truths to be self-evident; that all men are created equal; that they are endowed by their Creator with certain unalienable Rights, that among these are Life, Liberty, and the pursuit of Happiness." Unalienable—or inalienable, natural, human—rights cannot be taken away (alienated) from us, although of course they are sometimes violated. Human rights have been appealed to in all the major social movements of the twentieth century, including the women's movement, the civil rights movement, the farm workers' movement, the gay rights movement, and the patients' rights movement (in health care).

The idea of human rights is the single most powerful moral concept in making cross-cultural moral judgments about customs and laws.[1] As such, the notions of human rights and legal rights are distinct. Legal rights are simply those the law of a given society says one has, whereas human rights are those we have as humans, whether the law recognizes them or not.

Rights ethics applies to engineering in many ways. It provides a powerful foundation for the special ethical requirements in engineering and other professions.[2] Most engineering codes of ethics enjoin holding paramount the safety, health, and welfare of the public, a requirement that can be interpreted as having respect for the public's rights to life, rights not to be injured by dangerous products, rights to privacy, and rights to receive benefits through fair and honest exchanges in a free marketplace. In addition, the basic right to liberty implies a right to give informed consent to the risks accompanying technological products, an idea developed in Chapter 4.

In addition to human rights, there are *special moral rights*—rights held by particular individuals rather than by every human being. For example, engineers and their employers have special moral rights that arise from their respective roles and the contracts they make with each other. Special rights are grounded in human rights, however indirectly. Thus, contracts and other types of promises create special rights because people

---

[1] See, for example, Patrick Hayden, ed., *The Philosophy of Human Rights* (St. Paul, MN: Paragon House, 2001); and James W. Nickel, *Making Sense of Human Rights,* 2nd ed. (Malden, MA: Blackwell, 2007).

[2] Alan H. Goldman, *The Moral Foundations of Professional Ethics* (Totowa, NJ: Rowman and Littlefield, 1980); and Alan Gewirth, "Professional Ethics: The Separatist Thesis," *Ethics* 96 (1986), 287.

have human rights to liberty that are violated when the under-
standings and commitments specified in contracts and promises
are violated. And when consumers purchase products, there is an
implicit contract, based on an implicit understanding, that the
products will be safe and useful.

Rights ethics gets more complex as we ask which kinds of
rights exist—only liberty rights, or also welfare rights? *Liberty
rights* are rights to exercise our liberty, and they place duties
on other people not to interfere with our freedom. (The "not"
explains why they are also called negative rights.) *Welfare rights*
are rights to benefits needed for a decent human life, when we
cannot earn those benefits, perhaps because we are severely
handicapped, and when the community has them available. (As
a contrast to negative rights, they are sometimes called positive
rights.)

Most rights ethicists affirm that both liberty and welfare
human rights exist.[3] Indeed, they contend that liberty rights
imply at least some basic welfare rights. What, after all, is the
point of saying that we have rights to liberty if we are utterly
incapable of exercising liberty because, for example, we are
unable to obtain the basic necessities, such as jobs, worker compensation for serious injuries, and health care? Shifting to legal
rights, most Americans also support selective welfare rights,
including a guaranteed public education of kindergarten through
twelfth grade, Medicare and Medicaid, Social Security, and reasonable accommodations for persons with disabilities.

Another influential version of rights ethics, however, denies
there are welfare human rights. *Libertarians* believe that only
liberty rights exist; there are no welfare rights. John Locke
(1632–1704), who was the first philosopher to carefully articulate a rights ethics, is often interpreted as a libertarian.[4] Locke's
version of human rights ethics was highly individualistic. He
viewed rights primarily as entitlements that prevent other
people from meddling in our lives. The individualistic aspect of
Locke's thought is reflected in the contemporary political scene
in the Libertarian political party and outlook, with its emphases
on protecting private property, dismantling welfare systems,
and opposition to extensive government regulation of business
and the professions. Libertarians take a harsh view of taxes and

---

[3] For example, Ronald Dworkin, *Taking Rights Seriously* (Cambridge, MA:
Harvard University Press, 1978); Alan Gewirth, *Human Rights* (Chicago:
University of Chicago Press, 1982).

[4] John Locke, *Two Treatises of Government* (Cambridge: Cambridge
University Press, 1960). Milton Friedman, discussed in Chapter 1, was another
influential libertarian thinker. For a critique of libertarian views of property
see Liam Murphy and Thomas Nagel, *The Myth of Ownership: Taxes and
Justice* (New York: Oxford University Press, 2002).

government involvement beyond the bare minimum necessary for national defense, a legal system, and the preservation of free enterprise.

Locke thought of property as whatever we gained by "mixing our labor" with things—for example, coming to own lumber by going into the wilderness and cutting down a tree. Today, however, our understanding of property is far more complex. Many believe that property is largely what the law and government specify as to how we can acquire and use material things. Even so, Locke's followers tended to insist that property was sacrosanct and that governments continually intruded on property rights, particularly in the form of excessive taxation and regulation.

Finally, both libertarians and other rights ethicists can agree that few rights are absolute, in the sense of being unlimited and having no justifiable exceptions. When rights conflict with rights in practical situations, thereby creating ethical dilemmas, good judgment is required in arriving at reasonable solutions about how to reasonably balance the rights.

Duty Ethics

Rights and duties are typically correlated with each other. For example, our right to life places duties on others not to kill us, and our right to liberty places duties on others not to interfere with our freedom. *Duty ethics* reverses the order of priority by beginning with duties and deriving rights from them. Although the similarities between duty ethics and rights ethics are pronounced, historically they developed as distinct moral traditions.

Duty ethics says that right actions are those required by duties to respect the liberty or autonomy (self-determination) of individuals. One duty ethicist suggests the following list of important duties: "(1) Do not kill. (2) Do not cause pain. (3) Do not disable. (4) Do not deprive of freedom. (5) Do not deprive of pleasure. (6) Do not deceive. (7) Keep your promises. (8) Do not cheat. (9) Obey the law. (10) Do your duty [referring to work, family, and other special responsibilities]."[5]

How do we know that these are our duties? Immanuel Kant (1724–1804), the most famous duty ethicist, argued that all such specific duties derive from one fundamental duty to respect persons.[6] Persons deserve respect because they are moral agents—capable of recognizing and voluntarily responding to moral

---

[5] Bernard Gert, *Common Morality* (New York: Oxford University Press, 2004), 20.

[6] Immanuel Kant, *Groundwork of the Metaphysics of Morals,* in Immanuel Kant, *Practical Philosophy*, trans. Mary J. Gregor (New York: Cambridge University Press, 1996), 80.

duty (or, like children, they potentially have such capacities). *Autonomy*—moral self-determination or self-governance—means having the capacity to govern one's life in accordance with moral duties. Hence, respect for persons amounts to respect for their moral autonomy.

We ought always to treat persons as having their own rational aims, and not merely use them for our ends. Immorality occurs when we reduce other people to mere means to our ends and needs. Violent acts such as murder, rape, and torture are obvious ways of treating people as mere objects serving our own purposes. We also fail to respect persons if we fail to provide support for them when they are in desperate need, and we can help them at little inconvenience to ourselves. Some duties, then, are to refrain from interfering with a person's liberty, and some express requirements to help them when they are in need, thereby paralleling the distinction between liberty and positive rights. Of course we need to use one another as means all the time: Business partners, managers and engineers, and faculty and students use each other to obtain their personal and professional ends. Immorality occurs when we *merely* use persons as means to our goals, rather than as autonomous agents who have their own goals.

We also have duties to ourselves, for we too are rational and autonomous beings. As examples, Kant said we have a duty not to commit suicide, which would bring an end to a valuable life; we have duties to develop our talents, as part of unfolding our rational natures; and we should avoid harmful drugs that undermine our ability to exercise our rationality. Obviously, Kant's repeated appeal to the idea of rationality makes a number of assumptions about morally worthy aims. After beginning with the minimal idea of rationality as the capacity to obey moral principles, he built in a host of specific goals as part of what it means to be rational.

Kant emphasized that duties are universal: They apply equally to all persons. Here again, the idea is that valid principles of duty apply to all rationally autonomous beings, and hence valid duties will be such that we can envision everyone acting on them. This idea of universal principles is often compared to the Golden Rule: Do unto others as you would have them do unto you; or, in its negative version, Do not do unto others what you would not want them to do to you.[7]

Finally, Kant insisted that moral duties are "categorical imperatives." As imperatives, they are injunctions or commands that we impose on ourselves as well as other rational beings.

---

[7] Jeffrey Wattles, *The Golden Rule* (New York: Oxford University Press, 1996).

As categorical, they require us to do what is right *because* it is right, unconditionally and without special incentives attached. For example, we should be honest because honesty is required by duty; it is required by our basic duty to respect the autonomy of others, rather than to deceive and exploit them for our own selfish purposes. "Be honest!" says morality—not because doing so benefits us, but because honesty is our duty. Morality is not an "iffy" matter that concerns hypothetical (conditional) imperatives, such as "If you want to prosper, be honest." A businessperson who is honest solely because honesty pays—in terms of profits from customers who return and recommend their services, as well as from avoiding jail for dishonesty—fails to fully meet the requirements of morality. In this way, morality involves attention to motives and intentions, an idea also important in virtue ethics.

Kant's ideas of respect for autonomy, duties to ourselves, universal duties, and categorical imperatives have been highly influential. However, he made one large mistake. He thought that everyday principles of duty, such as "Do not lie" and "Keep your promises," were *absolute* in the sense of never having justifiable exceptions. In doing so, he conflated three ideas: (1) universality—moral rules apply to all rational agents; (2) categorical imperatives—moral rules command what is right because it is right; and (3) absolutism—moral rules have no justified exceptions. Nearly all ethicists reject Kant's absolutism, even ethicists who embrace his ideas of universality and categorical imperatives.

The problem with absolutism should be obvious. As we emphasized in Chapter 2, moral reasons are many and varied, including those expressed by principles of duty. Given the complexity of human life, they invariably come into conflict with each other, thereby creating moral dilemmas. Contemporary duty ethicists recognize that many moral dilemmas are resolvable only by recognizing some valid exceptions to simple principles of duty. Thus, engineers have a duty to maintain confidentiality about information owned by their corporations, but that duty can be overridden by the paramount duty to protect the safety, health, and welfare of the public.

To emphasize that most duties have some justified exceptions, philosophers now use the expression *prima facie duties*.[8] In this technical sense, prima facie simply means "might have justified exceptions" (rather than "at first glance"). Most duties are prima facie—they sometimes have permissible or obligatory exceptions. Indeed, the same is true of most rights and other moral prin-

---

[8] W. D. Ross, *The Right and the Good* (Oxford: Oxford University Press, 1946).

ciples, and hence today the term *prima facie* is also applied to rights and moral rules of all kinds.

Utilitarianism

Rights ethics and duty ethics agree that some types of actions, for example being fair and truthful, are (prima facie) obligatory for reasons independent of their consequences. In contrast, utilitarianism says the sole standard of right action is good consequences. There is only one general moral requirement: Produce the most good for the most people, giving equal consideration to everyone affected. (The word *utility* is sometimes used to refer to good consequences and other times to the balance of good over bad consequences.)

At first glance, the utilitarian standard seems simple and plausible. Surely morality involves producing good consequences—especially in engineering. Indeed, utilitarian modes of thinking are reflected in cost-benefit analyses: Tally up the likely good consequences of various options or proposals; do likewise for the likely bad consequences; and then favor that proposal which maximizes the overall good. Utilitarianism also seems a straightforward way to interpret the central principle in most engineering codes: "Engineers shall hold paramount the safety, health and welfare of the public in the performance of their professional duties." After all, *welfare* is a rough synonym for *overall good* (utility), and safety and health might be viewed as especially important aspects of that good.

Yet, what exactly is the good to be maximized? And should we maximize the good with respect to each situation, or instead with regard to general rules (policies, laws, principles in codes of ethics)? Depending on how these questions are answered, utilitarianism takes different forms.

To begin with, what is the standard for measuring *good* consequences? Specifically, what is *intrinsic good*—that is, good considered just by itself? All other good things are *instrumental goods* in that they provide means (instruments) for gaining intrinsic goods. Some utilitarians consider pleasure to be the only intrinsic good. But that seems counterintuitive—there is nothing good about the pleasures of tyrants and sadists take in inflicting suffering.

John Stuart Mill believed that happiness was the only intrinsic good, and hence he understood utilitarianism as the requirement to produce the greatest amount of happiness.[9] What is happiness? Sometimes Mill confused it with pleasures and enjoyments,

---

[9] John Stuart Mill, *Utilitarianism, with Critical Essays,* ed. Samuel Gorovitz (Indianapolis, IN: Bobbs-Merrill, 1971).

which are short-term, feel-good states of consciousness. In the main, however, Mill thought of happiness as (a) a life rich in pleasures, especially the "higher" pleasures of friendship, love, and intellectual endeavors, mixed with some inevitable pains, plus (b) a pattern of activities and relationships that we can affirm as the way we want our lives to be.

There are alternative theories of intrinsic good. Some utilitarians understand intrinsic goods as those which a reasonable person would pursue, or those which satisfy rational desires—those that we can affirm after fully examining them in light of relevant information, for example, love, friendship, appreciation of beauty, in addition to happiness. In sharp contrast, most economists adopt a preference theory: What is good is what individuals prefer, as manifested in what they choose to purchase in the marketplace. Arguments over which, if any, of these theories of intrinsic good obviously complicate utilitarian ethical theories.

In addition to developing a plausible view of intrinsic good, we need to decide whether to focus on individual actions or general rules. Classical, nineteenth-century utilitarians such as Mill believed in *act-utilitarianism:* A particular action is right if it is likely to produce the most good for the most people in a given situation, compared with alternative options available. The standard can be applied at any moment. Right now, should you continue reading this chapter? You might instead take a break, go to sleep, see a movie, or pursue any number of other options. Each option would have both immediate and long-term consequences that can be estimated. The right action is the one that produces the most overall good, taking into account everyone affected.

Yet, act-utilitarianism seems vulnerable to objections. It apparently permits some actions that we know, on other grounds, are patently immoral. Suppose that stealing a computer from my employer, an old one scheduled for replacement anyway, benefits me significantly and causes only miniscule harm to the employer and others. We know that the theft is unethical, and hence act-utilitarianism seems to justify wrongdoing. Again, suppose that in a particular situation more good is promoted by keeping the public ignorant about serious dangers, for example, by not informing them about a hidden fault in a car or building. Or suppose that it will improve company morale if several disliked engineers are fired after being blamed for mistakes they did not make. Doing so is unfair, but the overall good is promoted.

Such difficulties lead many, perhaps most, utilitarians to shift to an alternative version of utilitarianism that says we should maximize the good through following rules that maximize good consequences, rather than through isolated actions. According to this view, called *rule-utilitarianism,* right actions are those required by rules that produce the most good for the most people.

Because rules interact with each other, we need to consider a set of rules. Thus, we should seek to discover and act on an *optimal moral code*—that set of rules which maximizes the public good more than alternative codes would (or at least as much as alternatives).[10]

Rule-utilitarians have in mind society-wide rules, but the same idea applies to rules stated in engineering codes of ethics. Thus, an engineering code of ethics is justified in terms of its overall good consequences (compared to alternative codes), and so engineers should abide by it even when an exception might happen to be beneficial. For example, if codified rules forbidding bribes and deception are justified, then even if a particular bribe or deception is beneficial in some situations, one should still refrain from them.

Discussion Questions

1. Americans are sometimes criticized for being too individualistic, and in particular for approaching moral issues with too great an emphasis on rights. Although we said that rights and duties are usually correlated with each other, what difference (if any) do you think would occur if Jefferson had written, "We hold these truths to be self-evident; that all people are created equal; that they owe duties of respect to all other persons, and are owed these duties in return"?

2. What does the Golden Rule imply concerning how engineers and corporations should behave toward customers in designing and marketing products? As a focus, discuss whether crash-test information should be made available to customers concerning the possibly harmful side effects of a particular automobile. Does it matter whether the negative or positive version of the Golden Rule is used?

3. Cost-benefit analyses typically *reflect* utilitarian thinking, but too often they are slanted toward what is good for corporations, rather than the good for everyone affected, as utilitarians require. A cost-benefit analysis identifies the good and bad consequences of some action or policy, usually in terms of dollars.[11] It weighs the total positives against the total negatives, and then compares the results to similar tallies of the consequences of alternative actions or rules. In the following case, was Ford justified in relying exclusively on a cost-benefit analysis, or were

---

[10] Richard B. Brandt, *A Theory of the Good and the Right* (Oxford: Clarendon Press, 1979).

[11] Matthew D. Adler and Eric A. Posner, eds., *Cost-Benefit Analysis: Legal, Economic, and Philosophical Perspectives* (Chicago: University of Chicago Press, 2001).

there additional moral considerations that they should have used in deciding whether to improve the safety of the Pinto? What might rights ethicists and duty ethicists, as well as rule-utilitarians, say about the case?

For years, the Pinto was the largest-selling subcompact car in America. During the early stages of its development, crash-worthiness tests revealed that the Pinto could not sustain a front-end collision without the windshield breaking. A quick-fix solution was adopted: The drive train was moved backward. As a result, the differential was moved very close to the gas tank. Thus many gas tanks collapsed and exploded on rear-end collisions at low speeds. In 1977, Mark Dowie published an article in *Mother Jones* magazine that divulged the cost-benefit analysis developed by Ford Motor Company in 1971 to decide whether to add an $11 part per car that would greatly reduce injuries by protecting the vulnerable fuel tank—a tank that exploded in rear-end collisions under 5 miles per hour.[12] The $11 seems an insignificant expense, even adjusting to current dollars, but in fact it would make it far more difficult to market a car that was to be sold for no more than $2,000. Moreover, the costs of installing the part on 11 million cars and another 1.5 million light trucks added up. The cost of not installing the part and instead paying out costs for death and injuries from accidents was projected using a cost-benefit analysis. The analysis estimated the worth of a human life at about $200,000, a figure borrowed from the National Highway Traffic Safety Administration. The cost per non-death injury was $67,000. These figures were arrived at by adding together such costs as a typical worker's future earnings, hospital and mortuary costs, and legal fees. In addition, it was estimated that approximately 180 burn deaths and another 180 serious burn injuries would occur each year. Multiplying these numbers together, the annual costs for death and injury was $49.5 million, far less than the estimated $137 million for adding the part, let alone the lost revenue from trying to advertise a car for the uninviting figure of $2,011, or else reducing profit margins.

4. Present and defend your view concerning the relative strengths and weaknesses of the views of libertarian rights ethicists and those rights ethicists who believe in both liberty and welfare rights. In doing so, comment on why libertarianism is having considerable influence today, and yet why the Libertarian Party repeatedly cannot win widespread support for its goals to dis-

---

[12] Mark Dowie, "Pinto Madness," *Mother Jones* (September–October 1977). See also Douglas Birsch and John H. Fielder, eds., *The Ford Pinto Case* (Albany, NY: State University of New York Press, 1994).

mantle all welfare programs, such as guaranteed public education from kindergarten to twelfth grade and health care for the elderly and low-income families.

5. Apply act-utilitarianism and rule-utilitarianism in resolving the following moral problems. Do the two versions of utilitarianism lead to the same or different answers to the problems?

   a. George had a bad reaction to an illegal drug he accepted from friends at a party. He calls in sick the day after, and when he returns to work the following day he looks ill. His supervisor asks him why he is not feeling well. Is it morally permissible for George to lie by telling his supervisor that he had a bad reaction to some medicine his doctor prescribed for him?

   b. Jillian was aware of a recent company memo reminding employees that office supplies were for use at work only. Yet she knew that most of the other engineers in her division thought nothing about occasionally taking home notepads, pens, computer disks, and other office "incidentals." Her eight-year-old daughter had asked her for a company-inscribed ledger like the one she saw her carrying. The ledger costs less than $20, and Jillian recalls that she has probably used that much from her personal stationery supplies during the past year for work purposes. Is it all right for her to take home a ledger for her daughter without asking her supervisor for permission?

6. Can utilitarianism provide a moral justification for engineers who work for tobacco companies, for example, in designing cigarette-making machinery? In your answer take account of the following facts (and others you may be aware of).[13] Cigarettes kill more than 400,000 Americans each year, which is more than the combined deaths caused by alcohol and drug abuse, car accidents, homicide, suicide, and acquired immunodeficiency syndrome (AIDS). Cigarette companies do much good by providing jobs (Philip Morris employs more than 150,000 people worldwide), through taxes (more than $4 billion paid by Philip Morris in a typical year), and through philanthropy. Most new users of cigarettes in the United States are teenagers (younger than eighteen years of age). There is disagreement over just how addictive cigarettes are, but adults have some choice in deciding whether to continue using cigarettes, and they may choose to continue using for reasons beyond the addictive potential of nicotine.

---

[13] Roger Rosenblatt, "How Do Tobacco Executives Live with Themselves?" *New York Times Magazine*, March 20, 1994, 34–41, 55.

The preceding ethical theories placed the primary emphasis on right acts and moral rules. Other ethical theories shift the focus to the kinds of persons we should aspire to be and become. *Virtue ethics* focuses on good character, and *self-realization ethics* focuses on self-fulfillment.

3.2 Virtue Ethics, Self-Realization Ethics

Virtue Ethics

Character is the pattern of virtues (morally desirable features) and vices (morally undesirable features) in persons. *Virtues* are desirable habits or tendencies in action, commitment, motive, attitude, emotion, ways of reasoning, and ways of relating to others. *Vices* are morally undesirable habits or tendencies. The words *virtue* and *vice* sound a bit old-fashioned. Words for specific virtues, however, remain familiar, both in engineering and in everyday life—for example, competence, honesty, courage, fairness, loyalty, and humility. Words for specific vices are also familiar: incompetence, dishonesty, cowardice, unfairness, disloyalty, and arrogance.

Aristotle (384–322 BC) suggested that the moral virtues are habits of reaching a proper balance between extremes, whether in conduct, emotion, or desire.[14] Virtues are tendencies to find the reasonable (*golden*) mean between the extremes of too much (excess) and too little (deficiency) with regard to particular aspects of our lives. Thus, truthfulness is the appropriate middle ground (mean) between revealing all information, in violation of tact and confidentiality (excess), and being secretive or lacking in candor (deficiency) in dealing with truth. Again, courage is the mean between foolhardiness (the excess of rashness) and cowardice (the deficiency of self-control) in confronting dangers. The most important virtue is practical wisdom, that is, morally good judgment, which enables us to discern the mean for all the other virtues.

The Greek word *arete* translates as either "virtue" or "excellence," an etymological fact that reinforces our theme of ethics and excellence going together in engineering. The most comprehensive virtue of engineers is responsible professionalism. This umbrella virtue implies four (overlapping) categories of virtues: public well-being, professional competence, cooperative practices, and personal integrity.

Public-spirited virtues are focused on the good of clients and the wider public. The minimum virtue is nonmaleficence, that is, the tendency not to harm others intentionally. As Hippocrates

---

[14] Aristotle, *Ethics*, trans. J. A. K. Thomson and Hugh Tredennick (New York: Penguin, 1976).

reportedly said in connection with medicine, "Above all, do no harm." Engineering codes of professional conduct also call for beneficence, which is preventing or removing harm to others and, more positively, promoting the public safety, health, and welfare. Also important is a sense of community, manifested in faith and hope in the prospects for meaningful life within professional and public communities. Generosity, which means going beyond the minimum requirements in helping others, is shown by engineers who voluntarily give their time, talent, and money to their professional societies and local communities. Finally, justice within corporations, government, and economic practices is an essential virtue in the profession of engineering.

Proficiency virtues are the virtues of mastery of one's profession, in particular mastery of the technical skills that characterize good engineering practice. Following Aristotle, some thinkers regard these values as intellectual virtues rather than distinctly moral ones. As they contribute to sound engineering, however, they are morally desirable features. The most general proficiency virtue is competence: being well prepared for the jobs one undertakes. Another is diligence: alertness to dangers and careful attention to detail in performing tasks by, for example, avoiding the deficiency of laziness and the excess of the workaholic. Creativity is especially desirable within a rapidly changing technological society.

Teamwork virtues are those that are especially important in enabling professionals to work successfully with other people. They include collegiality, cooperativeness, loyalty, and respect for legitimate authority. Also important are leadership qualities that play key roles within authority-structured corporations, such as the responsible exercise of authority and the ability to motivate others to meet valuable goals.

Finally, *self-governance virtues* are those necessary in exercising moral responsibility.[15] Some of them center on moral understanding and perception: for example, self-understanding and good moral judgment—what Aristotle calls practical wisdom. Other self-governance virtues center on commitment and on putting understanding into action: for example, courage, self-discipline, perseverance, conscientiousness, fidelity to commitments, self-respect, and integrity. Honesty falls into both groups of self-governance virtues, for it implies both truthfulness in speech and belief and trustworthiness in commitments.

---

[15] John Kekes, *The Examined Life* (Lewisburg, PA: Bucknell University Press, 1988).

Virtue ethics takes contrasting forms, depending on which virtues are emphasized. To illustrate, we will contrast Samuel Florman's emphasis on conscientiousness and team-work virtues with Alasdair MacIntyre's emphasis on wider loyalties to community.

Samuel Florman is famous for his celebration of the existential pleasures of engineering, that is, the deeply rooted and elemental satisfactions in engineering that contribute to happiness.[16] These pleasures have many sources. There is the desire to improve the world, which engages an individual's sense of personal involvement and power. There is the challenge of practical and creative effort, including planning, designing, testing, producing, selling, constructing, and maintaining, all of which bring pride in achieving excellence in the technical aspects of one's work. There is the desire to understand the world—an understanding that brings wonder, peace, and sense of being at home in the universe. There is the sheer magnitude of natural phenomena—oceans, rivers, mountains, and prairies—that both inspires and challenges the design of immense ships, bridges, tunnels, communication links, and other vast undertakings. There is the presence of machines that can generate a comforting and absorbing sense of a manageable, controlled, and ordered world. Finally, engineers live with a sense of helping, of contributing to the well-being of other human beings.

In elaborating on these pleasures, Florman implicitly sets forth a virtue ethics. In his view, the essence of engineering ethics is captured by the word *conscientiousness,* which combines competence and loyalty.[17] "Competence" does not mean minimally adequate, but instead performing with requisite skill and experience. It implies exercising due care, persistence and diligence, honesty, attention to detail, and sometimes creativity. Loyalty means serving the interests of organizations that employ us, within the bounds of the law. Unlike libertarians who favor minimum government regulation, Florman places great emphasis on laws as setting the basic rules governing engineering. Within a democratic setting in which laws express a public consensus, economic competition among corporations makes possible technological achievements that benefit the public. Competition depends on engineers who are loyal to their organizations, rather

---

[16] Samuel C. Florman, *The Existential Pleasures of Engineering*, 2nd ed. (New York: St. Martin's Griffin, 1994).

[17] Samuel C. Florman, *The Civilized Engineer* (New York: St. Martin's Press, 1987), 101.

than engineers "filtering their everyday work through a sieve of ethical sensitivity."[18]

Whereas Florman defends the priority of duties to employers, most professional codes that require engineers to hold paramount the safety, health, and welfare of the public, which does seem to imply "filtering their everyday work through a sieve of ethical sensitivity." Such an emphasis on the good of community is found in Alasdair MacIntyre's emphasis on public-spirited virtues in thinking about professions.[19] MacIntyre conceives of professions as valuable social activities, which he calls *social practices.*

A social practice is "any coherent and complex form of socially established cooperative human activity through which goods internal to that form of activity are realized in the course of trying to achieve those standards of excellence which are appropriate to, and partially definitive of, that form of activity."[20] To clarify, *internal goods* are good things (products, activities, experiences, etc.) that are so essential to a social practice that they partly define it. In engineering these goods are safe and useful technological products—products that can be further specified with regard to each area of engineering. Additional internal goods are *personal goods* connected with meaningful work, such as personal meaning in working *as an engineer* to create useful and safe public goods and services. In contrast, *external goods* can be earned in or outside specific professions, such as money, power, self-esteem, and prestige. Although both internal and external goods are important, excessive concern for external goods, whether by individuals or organizations, threatens internal goods and undermines social practices.

Whether in Florman's or MacIntyre's version, virtue ethics seems vulnerable to the criticisms that it is incomplete and too vague. The meaning and requirements of virtues need to be spelled out in terms of at least rough guidelines or rules, lest the virtues fail to provide adequate moral guidance.[21] For example, honesty requires certain kinds of actions, done from certain kinds of motives. It implies a disposition, among other things, not to tell lies (without special justification) because lying disrespects persons and otherwise causes harm.

---

[18] Samuel C. Florman, *Blaming Technology* (New York: St. Martin's Press, 1981), 162–80.

[19] Some of the following material is from Mike W. Martin, "Personal Meaning and Ethics in Engineering," *Science and Engineering Ethics* 8 (2002): 545–60. It is used with permission of the publisher.

[20] Alasdair MacIntyre, *After Virtue*, 2nd ed. (South Bend, IN: University of Notre Dame Press, 1984), 187.

[21] James Rachels and Stuart Rachels, *The Elements of Moral Philosophy*, 5th ed. (Boston: McGraw-Hill, 2007), 173–90. In reply to the objection, see Rosalind Hursthouse, *On Virtue Ethics* (New York: Oxford University Press, 1999).

Self-Realization Ethics

Each of the preceding ethical theories leaves considerable room for self-interest, that is, for pursuing what is good for oneself. Thus, utilitarianism says that self-interest should enter into our calculations of the overall good; rights ethics says we have rights to pursue our legitimate interests; duty ethics says we have duties to ourselves; and virtue ethics links our personal good with participating in communities and social practices. *Self-realization ethics,* however, gives greater prominence to self-interest and to personal commitments that individuals develop in pursuing self-fulfillment. As with the other ethical theories, we will consider two versions, this time depending on how the self (the person) is conceived. In a community-oriented version, the self to be realized is understood in terms of caring relationships and communities. In a second version, called ethical egoism, the self is conceived in a highly individualistic manner.

The community-oriented version of self-realization ethics says that each individual ought to pursue self-realization, but it emphasizes the importance of caring relationships and communities in understanding self-realization. It emphasizes that we are social beings whose identities and meaning are linked to the communities in which we participate. This theme is expressed by F. H. Bradley (1826–1924): "The 'individual' apart from the community is an abstraction. It is not anything real, and hence not anything that we can realize. . . . I am myself by sharing with others."[22]

Individuals vary greatly in what they desire most strongly, and also in their talents and virtues.[23] Self-realization ethics points to the highly personal commitments that motivate, guide, and give meaning to the work of engineers and other professionals. These commitments enter into the core of an individual's character.[24] As such, they reflect what engineers care about deeply in ways that evoke their interest and energy, shape their identities, and generate pride or shame in their work. Personal commitments are commitments that are not incumbent on everyone—for example, specific humanitarian, environmental, religious, political, aesthetic, supererogatory, and family commitments. They also include, however, commitments to obligatory professional standards, especially when these are linked to an individual's broader value perspective.

---

[22] F. H. Bradley, *Ethical Studies* (New York: Oxford University Press, 1962), 173.

[23] Alan Gewirth, *Self-Fulfillment* (Princeton, NJ: Princeton University Press, 1998).

[24] Bernard Williams, "Persons, Character and Morality," in *Moral Luck* (New York: Cambridge University Press, 1981), 5.

Personal commitments are relevant in many ways to professional life, including one's choice of career and choice of jobs. Most important, they create meaning; thereby they motivate professionalism throughout long careers. Professions offer special opportunities for meaningful work, which explains much of their attraction to talented individuals. The relevant idea of meaning has subjective aspects—a "sense of meaning" that enlivens one's daily work and life. It also has objective aspects—the justified values that make work worthwhile and help make life worth living. In the following passage Joanne B. Ciulla has in mind both subjective and objective meaning:

> "Meaningful work, like a meaningful life, is morally worthy work undertaken in a morally worthy organization. Work has meaning *because* there is some good in it. The most meaningful jobs are those in which people directly help others or create products that make life better for people. Work makes life better if it helps others; alleviates suffering; eliminates difficult, dangerous, or tedious toil; makes someone healthier and happier; or aesthetically or intellectually enriches people and improves the environment in which we live."[25]

As just one illustration of personal commitments, and of the motivation and guidance they generate, consider the commitment to be creative, as illustrated by Jack Kilby who coinvented the microchip. The invention has had momentous importance in making possible the development of today's powerful computers, so much so that in 2000 Kilby was awarded a Nobel Prize—a rare event for an engineer, as Nobel Prizes are usually given for fundamental contributions to science, not engineering. In retrospect, the idea behind the microchip seems simple, as do many creative breakthroughs. During the 1950s the miniaturization of transistors was being pursued at a relentless pace, but it was clear there would soon be a limit to the vast number of minute components that could be wired together. Kilby was well aware of the problem, sensed the need for a fundamentally new approach, and was driven by personal commitments to be creative in finding a solution. In July 1958, only a few weeks after starting a new job at Texas Instruments, he discovered the solution: Make all parts of the circuit out of one material integrated on a piece of silicon, thereby removing the need to wire together miniature components.

In making his discovery, Kilby was not pursuing a grand intention to provide humanity with the remarkable goods the

---

[25] Joanne B. Ciulla, *The Working Life: The Promise and Betrayal of Modern Work* (New York: Times Books, 2000), 225–26. See also Arnold Pacey, *Meaning in Technology* (Cambridge, MA: MIT Press, 1999).

microchip would make possible, although it is true he was known for his everyday kindness to colleagues. When he was about to give his Nobel lecture, he was introduced as having made the invention that "launched the global digital revolution, making possible calculators, computers, digital cameras, pacemakers, the Internet, etc."[26] In response, he told a story borrowed from another Nobel laureate: "When I hear that kind of thing, it reminds me of what the beaver told the rabbit as they stood at the base of Hoover Dam: 'No, I didn't build it myself, but it's based on an idea of mine.'"

Was Kilby merely seeking money, power, fame, and other rewards just for himself? No, although these things mattered to him. As one biographer suggests, "we see nothing extraordinary in Jack Kilby's private ambition or in his aim to find personal fulfillment through professional achievement. In that regard he was the same as the rest of us: We all pick professions with a mind to fulfilling ourselves."[27] Primarily, Kilby was pursuing interests he had developed years earlier in how to solve technical problems in engineering. In this regard he was exceptional only in his passion for engineering work. Like many creative individuals, he was persistent to the point of being driven, and he found great joy in making discoveries. But even saying this by itself would be misleading. The accurate observation is that he had multiple motives, including motives to advance technology, to be compensated for his work, and to do some good for others.

Building on this observation, we might sort the motives of professionals into three categories: proficiency, compensation, and moral.

Proficiency motives, and their associated values, center on excellence in meeting the technical standards of a profession, together with related aesthetic values of beauty. The undergraduate curriculum for engineering is generally acknowledged to be more rigorous and difficult than the majority of academic disciplines. We might guess that students are attracted to engineering in part because of the challenge it offers to intelligent people. Do empirical studies back up this somewhat flattering portrayal? To a significant extent, yes. Typically, students are motivated to enter engineering primarily by a desire for interesting and challenging work. They have an "activist orientation" in the sense of wanting to create concrete objects and systems—to

---

[26] T. R. Reid, *The Chip* (New York: Random House, 2001), 265. Robert Noyce would no doubt have shared the prize had he not died 10 years earlier (the award is not given posthumously). See Leslie Berlin, *The Man Behind the Microchip: Robert Noyce and the Invention of Silicon Valley* (New York: Oxford University Press, 2005).

[27] Jeffrey Zygmont, *Microchip* (Cambridge, MA: Perseus Publishing, 2003), 3.

build them and to make them work. They are more skilled in math than average college students, although they tend to have a low tolerance for ambiguities and uncertainties that cannot be measured and translated into figures.[28]

Compensation motives are for social rewards such as income, power, recognition, and job or career stability. We tend to think of these motives and values as self-interested, and in a large degree they are. Yet most people seek money for additional reasons, such as to benefit family members or even to be able to help others in need. In addition, financial independence prevents one from becoming a burden on others. In general, due regard for one's self-interest is a moral virtue—the virtue of prudence—assuming it does not crowd out other virtues.

Moral motives include desires to meet one's responsibilities, respect the rights of others, and contribute to the well-being of others. Such motives of moral respect and caring involve affirming that other people have inherent moral worth. In addition, moral concern involves maintaining self-respect and integrity—valuing oneself as having equal moral worth and seeking to develop one's talents.

For the most part, these motives are interwoven and mutually supportive. All of them, not only moral motives, contribute to providing valuable services to the community, as well as professional relationships among engineers, other involved workers, and clients. Engineering is demanding, and it requires engineers to summon and to integrate a wide range of motivations. Indeed, life itself is demanding, and it can be argued that our survival requires constant interweaving and cross-fertilization of motives.[29]

For many engineers, we should add, moral motivation and commitments are interwoven with spiritual and religious ones. Here are two examples. Egbert Schuurman is a Dutch Calvinist engineer who has written extensively on technology. Highlighting the dangers of technology, he calls for redirecting technology to serve morally worthy aims, both human liberation and respect for the environment. He and his coauthors of *Responsible Technology* articulate normative principles for design. They include: cultural appropriateness (preserving valuable institutions and practices within a particular society); openness (divulging to the public the value judgments expressed in products and also their known effects); stewardship (frugality in the use of natural resources and energy); harmony (effectiveness of products together with

---

[28] Robert Perrucci and Joel E. Gerstl, *Profession without Community: Engineers in American Society* (New York: Random House, 1969), 27–52.
[29] Mary Midgley, *Beast and Man: The Roots of Human Nature*, rev. ed. (New York: New American Library, 1994), 331.

promoting social unity); justice (respect for persons); caring (for colleagues and workers); and trustworthiness (deserving consumers' trust).[30]

The second example is Mark Pesce, who invented dial-up networking. In 1994, Pesce and a colleague developed the Virtual Reality Modeling Language (VRML), which allowed three-dimensional models to be placed on the World Wide Web. Emphasizing the importance of spiritual attitudes in his work, he makes it clear that his beliefs are neither orthodox nor associated solely with any one world religion. He characterizes his beliefs as "a mélange of a lot of different religious traditions, including Christian, pre-Christian, Buddhist, Taoist and so on," integrated into a type of "paganism," which is "a practice of harmony, a religion of harmony with yourself and the environment."[31] He is aware that his contributions to technology can be used as tools of communication or weapons of domination. Spiritual attitudes seek ways to allow aspects of the sacred into technology, to find ways for technology to make human life more interconnected through global communication, as well as attuned to nature, and to allow individuals to express themselves in more broadly creative ways through the Web.

These examples barely hint at the myriad ways in which personal commitments, ideals, and meaning enter into professional ethics, including how individuals construe codified responsibilities.[32] Later we offer additional illustrations; for example, in Chapter 7 we comment on personal commitments in connection with whistle-blowing, and in Chapter 8 we comment on environmental commitments.

Ethical Egoism

Ethical egoism is a more individualistic version of self-realization ethics that says each of us ought always and only to promote our self-interest. The theory is *ethical* in that it is a theory about morality, and it is *egoistic* because it says the sole duty of each of us is to maximize our well-being. Self-interest is understood as our long-term and enlightened well-being (good, happiness), rather than a narrow, short-sighted pursuit of immediate plea-

---

[30] Stephen V. Monsma et al., *Responsible Technology* (Grand Rapids, MI: William B. Eerdmans Publishing, 1986), 171–77. See also Egbert Schuurman, "The Modern Babylon Culture," in *Technology and Responsibility,* ed. Paul Durbin (Dordrecht, Holland: D. Reidel, 1987).

[31] "Virtually Sacred," an interview in W. Mark Richardson and Gordy Slack, *Faith in Science* (London: Routledge, 2001), 104.

[32] For additional examples, see: Caroline Whitbeck, *Ethics in Engineering Practice and Research* (New York: Cambridge University Press, 1998), 306–12; and Mary Tiles and Hans Oberdiek, *Living in a Technological Culture: Human Tools and Human Values* (New York: Routledge, 1995), 172–75.

sures that leaves us frustrated or damaged in the long run. Thus, Thomas Hobbes (1588–1679) and Ayn Rand (1905–1982) recommended a "rational" concern for one's long-term interests. Nevertheless, ethical egoism sounds like an endorsement of selfishness. It implies that engineers should think first and last about what is beneficial to themselves, an implication at odds with the injunction to keep paramount the public health, safety, and welfare. As such, ethical egoism is an alarming view.

Are there any arguments to support ethical egoism? Rand offered three arguments. First, she emphasized the importance of self-respect, and then portrayed altruism toward others as incompatible with valuing oneself. She contended that acts of altruism are degrading, both to others and to oneself: "altruism permits no concept of a self-respecting, self-supporting man—a man who supports his life by his own effort and neither sacrifices himself nor others."[33] This argument contains one true premise: Independence is a value of great importance, especially in democratic and capitalistic economies. Yet, independence is not the only important value. In infancy, advanced age, and various junctures in between, each of us is vulnerable. We are also interdependent, as much as independent. Self-respect includes recognition of our vulnerabilities and interdependencies, and certainly it is compatible with caring about other persons as well as about ourselves.

Rand's second argument was that the world would be a better place if all or most people embraced ethical egoism. Especially in her novels, Rand portrayed heroic individuals who by pursuing their self-interest indirectly contribute to the good of others. She dramatized Adam Smith's "invisible hand" argument, set forth in 1776 in *The Wealth of Nations*. According to Smith, in the marketplace individuals do and should seek their own economic interests, but in doing so it is as if each businessperson were "led by an invisible hand to promote an end which was no part of his intention."[34] To be sure, Smith had in mind the invisible hand of God, whereas Rand was an atheist, but both appeal to the general good for society of self-seeking in the professions and business. This argument, too, contains an enormously important truth (although unrestrained capitalism does not always maximizes the general good). Nevertheless, contrary to Rand, this argument does not support ethical egoism. For notice that it assumes we ought to care about the well-being of others, for their sake—something denied by ethical egoism itself! And once

---

[33] Ayn Rand, *The Virtue of Selfishness* (New York: New American Library, 1964), ix, italics removed.

[34] Adam Smith, *An Inquiry into the Nature and Causes of the Wealth of Nations*, vol. 1 (New York: Oxford University Press, 1976), 456.

the general good becomes the moral touchstone, we are actually dealing with a version of utilitarianism.

Rand's third and most important argument was more complex, and it leads to a discussion of human nature and motivation. It asserted that ethical egoism was the only psychologically realistic ethical theory. By nature, human beings are exclusively self-seeking; our sole motives are to benefit ourselves. More fully, *psychological egoism* is true: All people are always and only motivated by what they believe is good for them in some respect. Psychological egoism is a theory about psychology, about what actually motivates human beings, whereas ethical egoism is a statement about how they ought to act. But if psychological egoism is true, ethical egoism becomes the only plausible ethical theory. If by nature we can only care about ourselves, we should at least adopt an enlightened view about how to promote our well-being.

Is psychological egoism true? Is the only thing an engineer or anyone else cares about, ultimately, their own well-being? Psychological egoism flies in the face of common sense, which discerns motives of human decency, compassion, and justice, as well as the proficiency motives mentioned earlier. It is difficult to refute psychological egoism directly, because it radically reinterprets both common sense and experimental data. But we can show that most arguments for psychological egoism are based on simple although seductive confusions. Here are four such arguments for psychological egoism.[35]

Argument 1. We always act on our own desires; therefore, we always and only seek something for ourselves, namely the satisfaction of our desires.

—In reply, the premise is true: We always act on our own desires. By definition, *my* actions are motivated by *my* desires together with *my* beliefs about how to satisfy those desires. But the conclusion does not follow. There are many different kinds of desires, depending on what the desire is for—the object of the desire. When we desire goods for ourselves, we are self-seeking; but when we desire goods for other people (for their sake), we are altruistic. The mere fact that in both instances we act on our own desires does nothing to support psychological egoism.

Argument 2. People always seek pleasures; therefore they always and only seek something for themselves, namely their pleasures.

---

[35] See James Rachels and Stuart Rachels, *The Elements of Moral Philosophy,* 68–88.

—In reply, there are different sources of pleasures. Taking pleasure in seeking and getting a good solely for oneself is different from taking pleasure in helping others.

Argument 3. We can always imagine there is an ulterior, exclusively self-seeking motive present whenever a person helps someone else; therefore people always and only seek goods for themselves.

—In reply, there is a difference between imagination and reality. We can imagine that people who help others are only seeking fame, but it does not follow that they actually are motivated in this way.

Argument 4. When we look closely, we invariably discover an element of self-interest in any given action; therefore people are solely motivated by self-interest.

—In reply, there is an enormous difference between the presence of an element of self-interest (asserted in the premise) and inferring the element is the only motive (asserted in the conclusion). Many actions have multiple motives, with an element of self-interest mixed in with concern for others.

We conclude that there are no sound reasons for believing psychological egoism, nor for believing ethical egoism. Nevertheless, having emphasized that self-seeking is not the only human motive, we certainly acknowledge that it is a very strong motive. Indeed, quite possibly *predominant egoism* is true: The strongest desire for most people most of the time is self-seeking.[36] It is also plausible to believe that most acts of helping and service to others involve mixed motives, that is, a combination of self-concern and concern for others.

Unlike psychological egoism, predominant egoism acknowledges human capacities for love, friendship, and community involvement. It also acknowledges engineers' capacities for genuinely caring about the public safety, health, and welfare. Engineers are strongly motivated by self-interest, but they are also capable of responding to moral reasons in their own right, as well as additional motives concerned with the particular nature of their work.

Which Ethical Theory Is Best?

Just as ethical theories are used to evaluate actions, rules, and character, ethical theories can themselves be evaluated. In this chapter, our concern has been to introduce some influential ethical theories rather than to try to determine which is preferable. Nevertheless, we argued against particular versions of each type

---

[36] Gregory S. Kavka, *Hobbesian Moral and Political Theory* (Princeton, NJ: Princeton University Press, 1986).

of theory. For example, we argued against act-utilitarianism, as compared with rule-utilitarianism, and we argued against ethical egoism. We hinted at our preference, as authors, for nonlibertarian versions of rights ethics. And we suggested that few duties are absolute, contrary to Kant.

Which criteria can be used in assessing ethical theories, and which criteria did we use? The criteria follow from the very definition of what ethical theories are. Ethical theories are attempts to provide clarity and consistency, systematic and comprehensive understanding, and helpful practical guidance in moral matters. Sound ethical theories succeed in meeting these aims.

First, sound ethical theories are clear and coherent. They rely on concepts (ideas) that are sufficiently clear to be applicable, and their various claims and principles are internally consistent.

Second, sound ethical theories organize basic moral values in a systematic and comprehensive way. They highlight important values and distinguish them from what is secondary. And they apply to all circumstances that interest us, not merely to a limited range of examples.

Third, and most important, sound ethical theories provide helpful guidance that is compatible with our most carefully considered moral convictions (judgments, intuitions) about concrete situations. Who does "our" refer to? It refers to each of us, in moral dialogue with others. To take an extreme case, if an ethical theory said it was all right for engineers to create extremely dangerous products without the public's informed consent, then that would show the theory is inadequate.

Of course, even our most carefully considered convictions can be mistaken, sometimes flagrantly so as with racists and other bigots. An important role of a sound ethical theory is to improve our moral insight into particular problems. Hence, there is an ongoing checking of an ethical theory (or general principles and rules) against the judgments about specific situations (cases, dilemmas, issues) that we are most confident are correct, and, in reverse, a checking of our judgments about specific situations by reference to the ethical theory. Theories and specific judgments are continually adjusted to each other in a back-and-forth process until we reach what John Rawls calls a *reflective equilibrium*.[37]

Which of the ethical theories most fully satisfies these criteria? In our view, some versions of rule-utilitarianism, rights ethics, duty ethics, virtue ethics, and self-realization ethics all satisfy the criteria in high degrees. We find ourselves more impressed

---

[37] John Rawls, *A Theory of Justice*, rev. ed. (Cambridge, MA: Harvard University Press, 1999), 18.

by the similarities and connections, rather than the differences, among the general types of theories.

Discussion Questions

1. Review the National Society of Professional Engineers (NSPE) Code of Ethics (in the Appendix). To what extent do its Preamble and Fundamental Canons rely on the language of the virtues? Try rewriting the Fundamental Canons entirely in terms of the virtues, and comment on what is lost or gained in doing so.

2. Discuss similarities and differences in the views of Samuel Florman and Alasdair MacIntyre, and what you find insightful and problematic in their views.

3. The following widely discussed case study was written by Bernard Williams (1929–2003). The case is about a chemist, but the issues it raises are equally relevant to engineering. What should George do to best preserve his integrity? Is it permissible for him to take the job and "compartmentalize" so as to separate his work and his personal commitments? In your answer, discuss whether in taking the job George would be compromising in either of the two senses of "compromise": (1) undermine integrity by violating one's fundamental moral principles; (2) settle moral dilemmas and differences by mutual concessions or to reconcile conflicts through adjustments in attitude and conduct.[38] What might rights ethicists and utilitarians say?

> George, who has just taken his doctorate in chemistry, finds it extremely difficult to get a job. He is not very robust in health, which cuts down the number of jobs he might be able to do satisfactorily. His wife has to go out to work to keep [i.e., to support] them, which itself causes a great deal of strain, since they have small children and there are severe problems about looking after them. The results of all this, especially on the children, are damaging. An older chemist, who knows about this situation, says that he can get George a decently paid job in a certain laboratory, which pursues research into chemical and biological warfare.[39]

4. With regard to each of the following cases, first discuss what morality requires and then what self-interest requires. Is the answer the same or different?

   a. Bill, a process engineer, learns from a former classmate who is now a regional compliance officer with the Occupational Safety and Health Administration (OSHA) that there will be

---

[38] Martin Benjamin, *Splitting the Difference: Compromise and Integrity in Ethics and Politics* (Lawrence, KS: University Press of Kansas, 1990).

[39] Bernard Williams, *Utilitarianism: For and Against* (New York: Cambridge University Press, 1973), 97–98.

an unannounced inspection of Bill's plant. Bill believes that unsafe practices are often tolerated in the plant, especially in the handling of toxic chemicals. Although there have been small spills, no serious accidents have occurred in the plant during the past few years. What should Bill do?[40]

b. On a midnight shift, a botched solution of sodium cyanide, a reactant in an organic synthesis, is temporarily stored in drums for reprocessing. Two weeks later, the day shift foreperson cannot find the drums. Roy, the plant manager, finds out that the batch has been illegally dumped into the sanitary sewer. He severely disciplines the night shift foreperson. Upon making discreet inquiries, he finds out that no apparent harm has resulted from the dumping.[41] Should Roy inform government authorities as is required by law in this kind of situation?

5. Wrongdoing by professionals is often caused in part by pressures within their organizations, but character remains important in understanding why only some professionals succumb to those pressures and engage in wrongdoing. Return to LeMessurier in the Citicorp case in Chapter 1 and discuss what kinds of character faults might tempt other engineers in his situation to simply ignore the problem. The faults might be general ones in an individual or those limited to the situation. Consider each of the following categories: (a) moral indifference and negligence, (b) intentional (knowing) wrongdoing, (c) professional incompetence, (d) bias or lack of objectivity, (e) fear, (f) lack of effort, (g) lack of imagination or perspective.

6. Discuss the following claim: "It is irrelevant what the motives of professionals are; what matters is that they do what is right." In your answer, distinguish questions about the motives for a specific right action and questions about habits or patterns of motivation throughout a career.

7. Long before H. G. Wells wrote *The Invisible Man,* Plato (428–348 BC), in *The Republic,* described a shepherd named Gyges who, according to a Greek legend, discovers a ring that enables him to become invisible when he turns its bezel. Gyges uses his magical powers to seduce the queen, kill the king, and take over an empire. If we have similar powers, why should we feel bound by moral constraints? In particular, if professionals are sufficiently powerful to pursue their desires without being caught for malfeasance, why should they care about the good of the wider public?

---

[40] Jay Matley, Richard Greene, and Celeste McCauley, "Health, Safety and Environment," *Chemical Engineering* 28 (September 1987), 115.
[41] Ibid., 117.

In your answer, reflect on the question "Why be moral?" Is the question asking for self-interested reasons for being moral, and if so does it already presuppose that only self-interest, not morality, provides valid reasons for conduct?

CHAPTER 4

Engineering as Social Experimentation

As it departed on its maiden voyage in April 1912, the *Titanic* was proclaimed the greatest engineering achievement ever. Not merely was it the largest ship the world had seen, having a length of almost three football fields; it was also the most glamorous of ocean liners, and it was touted as the first fully safe ship. Because the worst collision envisaged was at the juncture of two of its sixteen watertight compartments, and as it could float with any four compartments flooded, the *Titanic* was believed to be virtually unsinkable.

Buoyed by such confidence, the captain allowed the ship to sail full speed at night in an area frequented by icebergs, one of which tore a large gap in the ship's side, flooding five compartments. Time remained to evacuate the ship, but there were not enough lifeboats. Because British regulations then in effect did not foresee vessels of this size, only 825 places were required in lifeboats, sufficient for a mere one-quarter of the *Titanic's* capacity of 3,547 passengers and crew. No extra precautions had seemed necessary for an unsinkable ship. The result: 1,522 dead (drowned or frozen) out of the 2,227 on board for the Titanic's first trip.[1]

The *Titanic* remains a haunting image of technological complacency. So many products of technology present potential dangers that engineering should be regarded as an inherently risky activity. To underscore this fact and help to explore its ethical implications, we suggest that engineering should be viewed as an *experimental* process. It is not, of course, an experiment

[1] Walter Lord, *A Night to Remember* (New York: Holt, 1976); Wynn C. Wade, *The Titanic: End of a Dream* (New York: Penguin, 1980); Michael Davie, *The Titanic* (London: The Bodley Head, 1986).

conducted solely in a laboratory under controlled conditions. Rather, it is an experiment on a social scale involving human subjects.

4.1 Engineering as Experimentation

Experimentation is commonly recognized as playing an essential role in the design process. Preliminary tests or simulations are conducted from the time it is decided to convert a new engineering concept into its first rough design. Materials and processes are tried out, usually employing formal experimental techniques. Such tests serve as the basis for more detailed designs, which in turn are tested. At the production stage further tests are run, until a finished product evolves. The normal design process is thus iterative, carried out on trial designs with modifications being made on the basis of feedback information acquired from tests. Beyond those specific tests and experiments, however, each engineering project taken as a whole may be viewed as an experiment.

Similarities to Standard Experiments

Several features of virtually every kind of engineering practice combine to make it appropriate to view engineering projects as experiments. First, any project is carried out in partial ignorance. There are uncertainties in the abstract model used for the design calculations; there are uncertainties in the precise characteristics of the materials purchased; there are uncertainties in the precision of materials processing and fabrication; there are uncertainties about the nature of the stresses the finished product will encounter. Engineers do not have the luxury of waiting until all the relevant facts are in before commencing work. At some point, theoretical exploration and laboratory testing must be bypassed for the sake of moving ahead on a project. Indeed, one talent crucial to an engineer's success lies precisely in the ability to accomplish tasks safely with only a partial knowledge of scientific laws about nature and society.

Second, the final outcomes of engineering projects, like those of experiments, are generally uncertain. Often in engineering it is not even known what the possible outcomes may be, and great risks may attend even seemingly benign projects. A reservoir may do damage to a region's social fabric or to its ecosystem. It may not even serve its intended purpose if the dam leaks or breaks. An aqueduct may bring about a population explosion in a region where it is the only source of water, creating dependency and vulnerability without adequate safeguards. A jumbo airplane may bankrupt the small airline that bought it as a status symbol. A special-purpose fingerprint reader may find its main application in the identification and surveillance of dissidents by totalitarian regimes. A nuclear reactor, the scaled-up version of a successful smaller model, may exhibit unexpected problems that endanger the surrounding population, leading to its untimely shutdown at great cost to owner and consumers alike. In the past, a hair dryer may have exposed the user to lung damage from the asbestos insulation in its barrel.

Third, effective engineering relies on knowledge gained about products both before and after they leave the factory—knowledge needed for improving current products and creating better ones. That is, ongoing success in engineering depends on gaining new knowledge, as does ongoing success in experimentation. Monitoring is thus as essential to engineering as it is to experimentation in general. To monitor is to make periodic observations and tests to check for both successful performance and unintended side effects. But as the ultimate test of a product's efficiency, safety, cost-effectiveness, environmental impact, and aesthetic value lies in how well that product functions within society, monitoring cannot be restricted to the in-house development or testing phases of an engineering venture. It also extends to the stage of client use. Just as in experimentation, both the intermediate and final results of an engineering project deserve analysis if the correct lessons are to be learned from it.

Learning from the Past

Usually engineers learn from their own earlier design and operating results, as well as from those of other engineers, but unfortunately that is not always the case. Lack of established channels of communication, misplaced pride in not asking for information, embarrassment at failure or fear of litigation, and plain neglect often impede the flow of such information and lead to many repetitions of past mistakes. Here are a few examples:

1. The *Titanic* lacked a sufficient number of lifeboats decades after most of the passengers and crew on the steamship *Arctic* had perished because of the same problem.

2. "Complete lack of protection against impact by shipping caused Sweden's worst ever bridge collapse on Friday as a result of which eight people were killed." Thus reported the *New Civil Engineer* on January 24, 1980. Engineers now recommend the use of floating concrete bumpers that can deflect ships, but that recommendation is rarely heeded as seen by the 1993 collapse of the Bayou Canot bridge that cost 43 passengers of the *Sunset Limited* their lives.

3. Valves are notorious for being among the least reliable components of hydraulic systems. It was a pressure relief valve, and a lack of definitive information regarding its open or shut state,

which contributed to the nuclear reactor accident at Three Mile Island on March 28, 1979. Similar malfunctions had occurred with identical valves on nuclear reactors at other locations. The required reports had been filed with Babcock and Wilcox, the reactor's manufacturer, but no attention had been given to them.[2]

These examples illustrate why it is not enough for engineers to rely on handbooks and computer programs without knowing the limits of the tables and algorithms underlying their favorite tools. They do well to visit shop floors and construction sites to learn from workers and testers how well the customers' wishes were met. The art of back-of-the-envelope calculations to obtain ballpark values with which to quickly check lengthy and complicated computational procedures must not be lost. Engineering demands practitioners who remain alert and well informed at every stage of a project's history and who exchange ideas freely with colleagues in related departments.

Contrasts with Standard Experiments

To be sure, engineering differs in some respects from standard experimentation. Some of those very differences help to highlight the engineer's special responsibilities. Exploring the differences can also aid our thinking about the moral responsibilities of all those engaged in engineering.

Experimental Control. One great difference arises with experimental control. In a standard experiment this involves the selection, at random, of members for two different groups. The members of one group receive the special, experimental treatment. Members of the other group, called the control group, do not receive that special treatment, although they are subjected to the same environment as the first group in every other respect.

In engineering, this is not the usual practice—unless the project is confined to laboratory experimentation—because the experimental subjects are human beings or finished and sold products out of the experimenter's control. Indeed, clients and consumers exercise most of the control because it is they who choose the product or item they wish to use. This makes it impossible to obtain a random selection of participants from various groups. Nor can parallel control groups be established based on random sampling. Thus it is not possible to study the effects that

---

[2] Robert Sugarman, "Nuclear Power and the Public Risk," *IEEE Spectrum* 16 (November 1979): 72.

changes in variables have on two or more comparison groups, and one must simply work with the available historical and retrospective data about various groups that use the product.

This suggests that the view of engineering as social experimentation involves a somewhat extended usage of the concept of experimentation. Nevertheless, "engineering as social experimentation" should not be dismissed as a mere metaphor. There are other fields where it is not uncommon to speak of experiments whose original purpose was not experimental in nature and that involve no control groups.

For example, social scientists monitor and collect data on differences and similarities between existing educational systems that were not initially set up as systematic experiments. In doing so they regard the current diversity of systems as constituting what has been called a *natural experiment* (as opposed to a deliberately initiated one).[3] Similarly, we think that engineering can be appropriately viewed as just such a natural experiment using human subjects.

Informed Consent. Viewing engineering as an experiment on a societal scale places the focus where it should be—on the human beings affected by technology, for the experiment is performed on persons, not on inanimate objects. In this respect, albeit on a much larger scale, engineering closely parallels medical testing of new drugs or procedures on human subjects.

Society has recently come to recognize the primacy of the subject's safety and freedom of choice as to whether to participate in medical experiments. Ever since the revelations of prison and concentration camp horrors in the name of medicine, an increasing number of moral and legal safeguards have arisen to ensure that subjects in experiments participate on the basis of informed consent (as discussed in Chapter 7).

Although current medical practice has increasingly tended to accept as fundamental the subject's moral and legal rights to give informed consent before participating in an experiment, contemporary engineering practice is only beginning to recognize those rights. We believe that the problem of informed consent, which is so vital to the concept of a properly conducted experiment involving human subjects, should be the keystone in the interaction between engineers and the public. We are talking about the lay public. When a manufacturer sells a new device to a knowledgeable firm that has its own engineering staff, there is usually an

---

[3] Alice M. Rivlin, *Systematic Thinking for Social Action* (Washington, DC: The Brookings Institution, 1971), 70.

agreement regarding the shared risks and benefits of trying out the technological innovation.

Informed consent is understood as including two main elements: knowledge and voluntariness. First, subjects should be given not only the information they request, but all the information needed to make a reasonable decision. Second, subjects must enter into the experiment without being subjected to force, fraud, or deception. Respect for the fundamental rights of dissenting minorities and compensation for harmful effects are taken for granted here.

The mere purchase of a product does not constitute informed consent, any more than does the act of showing up on the occasion of a medical examination. The public and clients must be given information about the practical risks and benefits of the process or product in terms they can understand. Supplying complete information is neither necessary nor in most cases possible. In both medicine and engineering there may be an enormous gap between the experimenter's and the subject's understanding of the complexities of an experiment. But whereas this gap most likely cannot be closed, it should be possible to convey all pertinent information needed for making a reasonable decision on whether to participate.

We do not propose a process resembling the preparation and release of environmental impact reports. Those reports should be carried out anyway when large projects are involved. We favor the kind of sound advice a responsible physician gives a patient when prescribing a course of drug treatment that has possible side effects. The physician must search beyond the typical sales brochures from drug manufacturers for adequate information; hospital management must allow the physician the freedom to undertake different treatments for different patients, as each case may constitute a different "experiment" involving different circumstances; finally, the patient must be readied to receive the information.

Likewise, engineers cannot succeed in providing essential information about a project or product unless there is cooperation by superiors and also receptivity on the part of those who should have the information. Management is often understandably reluctant to provide more information than current laws require, fearing disclosure to potential competitors and exposure to potential lawsuits. Moreover, it is possible that, paralleling the experience in medicine, clients or the public may not be interested in all of the relevant information about an engineering project, at least not until a crisis looms. It is important nevertheless that all avenues for disseminating such information be kept open and ready.

We note that the matter of informed consent is surfacing indirectly in the debate over acceptable forms of energy, intensified in light of the need to confront global warming. Representatives of the nuclear industry can be heard expressing their impatience with critics who worry about reactor malfunction while engaging in statistically more hazardous activities such as driving automobiles and smoking cigarettes. But what is being overlooked by those industry representatives is the common enough human readiness to accept *voluntarily undertaken risks* (as in daring sports), even while objecting to *involuntary risks* resulting from activities in which the individual is neither a direct participant nor a decision maker. In other words, we all prefer to be the subjects of our own experiments rather than those of somebody else. When it comes to approving a nearby oil-drilling platform or a nuclear plant, affected parties expect their consent to be sought no less than it is when a doctor contemplates surgery.

Prior consultation of the kind suggested can be effective. When Northern States Power Company (Minnesota) was planning a new power plant, it got in touch with local citizens and environmental groups before it committed large sums of money to preliminary design studies. The company was able to present convincing evidence regarding the need for a new plant and then suggested several sites. Citizen groups responded with a site proposal of their own. The latter was found acceptable by the company. Thus, informed consent was sought from and voluntarily given by those the project affected, and the acrimonious and protracted battles so common in other cases where a company has already invested heavily in decisions based on engineering studies alone was avoided.[4] Note that the utility company interacted with groups that could serve as proxy for various segments of the rate-paying public. Obviously it would have been difficult to involve the rate-payers individually.

We endorse a broad notion of informed consent, or what some would call *valid consent* defined by the following conditions:[5]

1. The consent was given voluntarily.
2. The consent was based on the information that a rational person would want, together with any other information requested, presented to them in understandable form.

[4] Peter Borrelli et al., *People, Power and Pollution* (Pasadena, CA: Environmental Quality Lab, California Institute of Technology, 1971), 36–39.
[5] Charles M. Culver and Bernard Gert, "Valid Consent," in *Conceptual and Ethical Problems in Medicine and Psychiatry*, ed. Charles M. Culver and Bernard Gert (New York: Oxford University Press, 1982).

3. The consenter was competent (not too young or mentally ill, for instance) to process the information and make rational decisions.

We suggest two requirements for situations in which the subject cannot be readily identified as an individual:

4. Information that a rational person would need, stated in understandable form, has been widely disseminated.

5. The subject's consent was offered in proxy by a group that collectively represents many subjects of like interests, concerns, and exposure to risk.

Knowledge Gained. Scientific experiments are conducted to gain new knowledge, whereas "engineering projects are experiments that are not necessarily designed to produce very much knowledge," according to a valuable interpretation of the social-experimentation paradigm by Taft Broome.[6] When we carry out an engineering activity as if it were an experiment, we are primarily preparing ourselves for unexpected outcomes. The best outcome in this sense is one that tells us nothing new but merely affirms that we are right about something. Unexpected outcomes send us on a search for new knowledge—possibly involving an experiment of the first (scientific) type. For the purposes of our model the distinction is not vital because we are concerned about the manner in which the experiment is conducted, such as that valid consent of human subjects is sought, safety measures are taken, and means exist for terminating the experiment at any time and providing all participants a safe exit.

Discussion Questions

1. On June 5, 1976, Idaho's Teton Dam collapsed, killing eleven people and causing $400 million in damage. The Bureau of Reclamation, which built the ill-fated Teton Dam, allowed it to be filled rapidly, thus failing to provide sufficient time to monitor for the presence of leaks in a project constructed with less-than-ideal soil.[7] Drawing on the concept of engineering as social experimentation, discuss the following facts uncovered by the General Accounting Office and reported in the press.

---

[6] Taft H. Broome Jr., "Engineering Responsibility for Hazardous Technologies," *Journal of Professional Issues in Engineering* 113 (April 1987): 139–49.

[7] Gaylord Shaw, "Bureau of Reclamation Harshly Criticized in New Report on Teton Dam Collapse," *Los Angeles Times*, June 4, 1977, part I: 3; Philip M. Boffey, "Teton Dam Verdict: Foul-up by the Engineers," *Science* 195 (January 1977): 270–72.

a. Because of the designers' confidence in the basic design of Teton Dam, it was believed that no significant water seepage would occur. Thus sufficient instrumentation to detect water erosion was not installed.

b. Significant information suggesting the possibility of water seepage was acquired at the dam site six weeks before the collapse. The information was sent through routine channels from the project supervisors to the designers and arrived at the designers the day after the collapse.

c. During the important stage of filling the reservoir, there was no around-the-clock observation of the dam. As a result, the leak was detected only five hours before the collapse. Even then, the main outlet could not be opened to prevent the collapse because a contractor was behind schedule in completing the outlet structure.

d. Ten years earlier the Bureau's Fontenelle Dam had experienced massive leaks that caused a partial collapse, an experience the bureau could have drawn on.

2. Research the collapse of the Interstate 35W Bridge in Minneapolis on August 1, 2007, which killed 13 people and injured 100 more. In light of the social experimentation model, discuss its causes and whether it could have been prevented.

3. Debates over responsibility for safety in regard to technological products often turn on who should be considered mainly responsible, the consumer ("buyer beware") or the manufacturer ("seller beware"). How might an emphasis on the idea of informed consent influence thinking about this question?

4. Thought models often influence thinking by effectively organizing and guiding reflection and crystallizing attitudes. Yet they usually have limitations and can themselves be misleading to some degree. With this in mind, critically assess the strengths and weaknesses you see in the social experimentation model.

One possible criticism you might consider is whether the model focuses too much on the creation of new products, whereas a great deal of engineering involves the routine application of results from past work and projects. Another point to consider is how informed consent is to be measured in situations where different groups are involved, as in the construction of a garbage incinerator near a community of people having mixed views about the advisability of constructing the incinerator.

4.2 Engineers as Responsible Experimenters

What are the responsibilities of engineers to society? Viewing engineering as social experimentation does not by itself answer this question. Although engineers are the main technical enablers or facilitators, they are far from being the sole experimenters. Their responsibility is shared with management, the public, and others. Yet their expertise places them in a unique position to monitor projects, to identify risks, and to provide clients and the public with the information needed to make reasonable decisions.

From the perspective of engineering as social experimentation, four features characterize what it means to be a responsible person while acting as an engineer: a conscientious commitment to live by moral values, a comprehensive perspective, autonomy, and accountability.[8] Or, stated in greater detail as applied to engineering projects conceived as social experiments:

1. A primary obligation to protect the safety of human subjects and respect their right of consent
2. A constant awareness of the experimental nature of any project, imaginative forecasting of its possible side effects, and a reasonable effort to monitor them
3. Autonomous, personal involvement in all steps of a project
4. Accepting accountability for the results of a project

Conscientiousness

People act responsibly to the extent that they conscientiously commit themselves to live according to moral values, instead of a consuming preoccupation with a narrowly conceived self-interest. By conscientious moral commitment we mean sensitivity to the full range of moral values and responsibilities relevant to a given situation, and the willingness to develop the skill and expend the effort needed to reach a reasonable balance among those considerations. Conscientiousness implies consciousness: open eyes, open ears, and an open mind.

The contemporary working conditions of engineers tend to narrow moral vision solely to the obligations that accompany employee status. More than 90 percent of engineers are salaried employees, most of whom work within large bureaucracies under great pressure to function smoothly within the organization. There are obvious benefits in terms of prudent self-interest and concern for one's family that make it easy to emphasize as primary the obligations to one's employer. Gradually the minimal negative duties, such as not falsifying data, not violating patent rights, and not breaching confidentiality, may come to be viewed as the full extent of moral aspiration.

Conceiving engineering as social experimentation restores the vision of engineers as guardians of the public interest, whose

---

[8] Graham Haydon, "On Being Responsible," *Philosophical Quarterly* 28 (1978): 46–57.

professional duty it is to hold paramount the safety, health, and welfare of those affected by engineering projects. And this helps to ensure that such safety and welfare will not be disregarded in the quest for new knowledge, the rush for profits, a narrow adherence to rules, or a concern over benefits for the many that ignores harm to the few.

The role of social guardian should not suggest that engineers force, paternalistically, their own views of the social good on society. For, as with medical experimentation on humans, the social experimentation involved in engineering should be restricted by the participant's voluntary and informed consent.

Comprehensive Perspective

Conscientiousness is blind without relevant factual information. Hence showing moral concern involves a commitment to obtain and properly assess all available information that is pertinent to meeting moral obligations. This means, as a first step, fully grasping the context of one's work, which makes it count as an activity having a moral import.

For example, in designing a heat exchanger, if I ignore the fact that it will be used in the manufacture of a potent, illegal hallucinogen, I am showing a lack of moral concern. It is this requirement that one be aware of the wider implications of one's work that makes participation in, say, a design project for a superweapon morally problematic—and that makes it sometimes convenient for engineers self-deceivingly to ignore the wider context of their activities, a context that may rest uneasily with conscience.

Another way of blurring the context of one's work results from the ever-increasing specialization and division of labor that makes it easy to think of someone else in the organization as responsible for what otherwise might be a bothersome personal problem. For example, a company may produce items with obsolescence built into them, or the items might promote unnecessary energy usage. It is easy to place the burden on the sales department: "Let them inform the customers—if the customers ask." It may be natural to thus rationalize one's neglect of safety or cost considerations, but it shows no moral concern. More convenient is a shifting of the burden to the government and voters: "We will attend to this when the government sets standards so our competitors must follow suit," or "Let the voters decide on the use of superweapons; we just build them."

These ways of losing perspective on the nature of one's work also hinder acquiring a full perspective along a second dimension of factual information: the consequences of what one does. And so although regarding engineering as social experimentation points out the importance of context, it also urges the engineer to view his or her specialized activities in a project as part of a larger whole having a social impact—an impact that may involve a variety of unintended effects. Accordingly, it emphasizes the need for wide training in disciplines related to engineering and its results, as well as the need for a constant effort to imaginatively foresee dangers.

It might be said that the goal is to practice "preventive technology," which parallels the idea of preventive medicine: The solution to the problem is not in successive cures to successive science-caused problems; it is in their prevention."[9] No amount of disciplined and imaginative foresight, however, can anticipate all dangers. Because engineering projects are inherently experimental in nature, they need to be monitored on an ongoing basis from the time they are put into effect. Individual practitioners cannot privately conduct full-blown environmental and social impact studies, but they can choose to make the extra effort needed to keep in touch with the course of a project after it has officially left their hands. This is a mark of personal identification with one's work, a notion that leads to the next aspect of moral responsibility.

Moral Autonomy

People are morally autonomous when their moral conduct and principles of action are their own, in a special sense derived from Kant: Moral beliefs and attitudes should be held on the basis of critical reflection rather than passive adoption of the particular conventions of one's society, church, or profession. This is often what is meant by "authenticity" in one's commitment to moral values. Those beliefs and attitudes, moreover, must be integrated into the core of an individual's personality in a manner that leads to committed action.

It is a comfortable illusion to think that in working for an employer, and thereby performing acts directly serving a company's interests, one is no longer morally and personally identified with one's actions. Selling one's labor and skills may make it seem that one has thereby disowned and forfeited power over one's actions.[10]

Viewing engineering as social experimentation can help overcome this tendency and restore a sense of autonomous participation in one's work. As an experimenter, an engineer is exercising the sophisticated training that forms the core of his or her identity as a professional. Moreover, viewing an engineering project

---

[9] Ruth M. Davis, "Preventative Technology: A Cure for Specific Ills," *Science* 188 (April 1975): 213.

[10] Elizabeth Wolgast, *Ethics of an Artificial Person: Lost Responsibility in Professions and Organizations* (Stanford, CA: Stanford University Press, 1992).

as an experiment that can result in unknown consequences should help inspire a critical and questioning attitude about the adequacy of current economic and safety standards. This also can lead to a greater sense of personal involvement with one's work.

The attitude of management plays a decisive role in how much moral autonomy engineers feel they have. It would be in the long-term interest of a high-technology firm to grant its engineers a great deal of latitude in exercising their professional judgment on moral issues relevant to their jobs (and, indeed, on technical issues as well). But the yardsticks by which a manager's performance is judged on a quarterly or yearly basis often discourage this. This is particularly true in our age of conglomerates, when near-term profitability is more important than consistent quality and long-term retention of satisfied customers.

Accountability

Finally, responsible people accept moral responsibility for their actions. Too often "accountable" is understood in the overly narrow sense of being culpable and blameworthy for misdeeds. But the term more properly refers to the general disposition of being willing to submit one's actions to moral scrutiny and be open and responsive to the assessments of others. It involves willingness to present morally cogent reasons for one's conduct when called on to do so in appropriate circumstances.

Submission to an employer's authority, or any authority for that matter, creates in many people a narrowed sense of accountability for the consequences of their actions. This was documented by some famous experiments conducted by Stanley Milgram during the 1960s.[11] Subjects would come to a laboratory believing they were to participate in a memory and learning test. In one variation, two other people were involved, the "experimenter" and the "learner." The experimenter was regarded by the subject as an authority figure, representing the scientific community. He or she would give the subject orders to administer electric shocks to the "learner" whenever the latter failed in the memory test. The subject was told the shocks were to be increased in magnitude with each memory failure. All this, however, was a deception. There were no real shocks, and the apparent "learner" and the "experimenter" were merely acting parts in a ruse designed to see how far the unknowing experimental subject was willing to go in following orders from an authority figure.

The results were astounding. When the subjects were placed in an adjoining room separated from the "learner" by a shaded glass window, more than half were willing to follow orders to the full

---

[11] Stanley Milgram, *Obedience to Authority* (New York: Harper & Row, 1974).

extent: giving the maximum electric jolt of 450 volts. This was in spite of seeing the "learner," who was strapped in a chair, writhing in (apparent) agony. The same results occurred when the subjects were allowed to hear the (apparently) pained screams and protests of the "learner," screams and protests that became intense from 130 volts on. There was a striking difference, however, when subjects were placed in the same room within touching distance of the "learner." Then the number of subjects willing to continue to the maximum shock dropped by one-half.

Milgram explained these results by citing a strong psychological tendency to be willing to abandon personal accountability when placed under authority. He saw his subjects ascribing all initiative, and thereby all accountability, to what they viewed as legitimate authority. And he noted that the closer the physical proximity, the more difficult it becomes to divest oneself of personal accountability.

The divorce between causal influence and moral accountability is common in business and the professions, and engineering is no exception. Such a psychological schism is encouraged by several prominent features of contemporary engineering practice.

First, large-scale engineering projects involve fragmentation of work. Each person makes only a small contribution to something much larger. Moreover, the final product is often physically removed from one's immediate workplace, creating the kind of "distancing" that Milgram identified as encouraging a lessened sense of personal accountability.

Second, corresponding to the fragmentation of work is a vast diffusion of accountability within large institutions. The often massive bureaucracies within which so many engineers work are bound to diffuse and delimit areas of personal accountability within hierarchies of authority.

Third, there is often pressure to move on to a new project before the current one has been operating long enough to be observed carefully. This promotes a sense of being accountable only for meeting schedules.

Fourth, the contagion of malpractice suits currently afflicting the medical profession is carrying over into engineering. With this comes a crippling preoccupation with legalities, a preoccupation that makes one wary of becoming morally involved in matters beyond one's strictly defined institutional role.

We do not mean to underestimate the very real difficulties these conditions pose for engineers who seek to act as morally accountable people on their jobs. Much less do we wish to say engineers are blameworthy for all the harmful side effects of the projects they work on, even though they partially cause those effects simply by working on the projects. That would be to confuse accountability with *blameworthiness,* and also to confuse *causal* responsibility with *moral* responsibility. But we do claim that engineers who endorse the perspective of engineering as a social experiment will find it more difficult to divorce themselves psychologically from personal responsibility for their work. Such an attitude will deepen their awareness of how engineers daily cooperate in a risky enterprise in which they exercise their personal expertise toward goals they are especially qualified to attain, and for which they are also accountable.

A Balanced Outlook on Law

Hammurabi, as king of Babylon, was concerned with strict order in his realm, and he decided that the builders of his time should also be governed by his laws. In 1758 BCE he decreed: "If a builder has built a house for a man and has not made his work sound, and the house which he has built has fallen down and so caused the death of the householder, that builder shall be put to death . . . If a builder has built a house for a man and does not make his work perfect and the wall bulges, that builder shall put that wall into sound condition at his own cost."[12] What should be the role of law in engineering, as viewed within our model of social experimentation?

The legal regulations that apply to engineering and other professions are becoming more numerous and more specific all the time. We hear many complaints about this trend, and a major effort to deregulate various spheres of our lives is currently under way. Nevertheless, we continue to hear cries of "there ought to be a law" whenever a crisis occurs or a special interest is threatened. This should not be surprising to us in the United States. We pride ourselves on being a nation that lives under the rule of law. We even delegate many of our decisions on ethical issues to an interpretation of laws. And yet this emphasis on law can cause problems in regard to ethical conduct quite aside from the more practical issues usually cited by those who favor deregulation.

For example, one of the greatest moral problems in engineering, and one fostered by the very existence of minutely detailed rules, is that of *minimal compliance*. This can find its expression when companies or individuals search for loopholes in the law that will allow them to barely keep to its letter even while violating its spirit. Or, hard-pressed engineers find it convenient to refer to standards with ready-made specifications as a substitute for original thought, perpetuating the "handbook mentality" and the repetition of mistakes. Minimal compliance led to the tragedy of the *Titanic:* Why should that ship have been equipped with

---

[12] Hammurabi, *The Code of Hammurabi*, trans. R. F. Harper (University of Chicago Press, 1904).

enough lifeboats to accommodate all its passengers and crew when British regulations at the time required only a lower minimum, albeit with smaller ships in mind?

Yet, remedying the situation by continually updating laws or regulations with further specifications may also be counterproductive. Not only will the law inevitably lag behind changes in technology and produce a judicial vacuum, there is also the danger of overburdening the rules and the regulators.[13]

Lawmakers cannot be expected always to keep up with technological development. Nor would we necessarily want to see laws changed with each innovation. Instead we empower rule-making and inspection agencies to fill the void. The Food and Drug Administration (FDA), Federal Aviation Agency (FAA), and the Environmental Protection Agency (EPA) are examples of these in the United States. Although they are nominally independent in that they belong neither to the judicial nor the executive branches of government, their rules have, for all practical purposes, the effect of law, but they are headed by political appointees.

Industry tends to complain that excessive restrictions are imposed on it by regulatory agencies. But one needs to reflect on why regulations may have been necessary in the first place. Take, for example, the U.S. Consumer Product Safety Commission's rule for baby cribs, which specifies that "the distance between components (such as slats, spindles, crib rods, and corner posts) shall not be greater than $2^3/_8$ inches at any point." This rule came about because some manufacturers of baby furniture had neglected to consider the danger of babies strangling in cribs or had neglected to measure the size of babies' heads.[14]

Again, why must regulations be so specific when broad statements would appear to make more sense? When the EPA adopted rules for asbestos emissions in 1971, it was recognized that strict numerical standards would be impossible to promulgate. Asbestos dispersal and intake, for example, are difficult to measure in the field. So, being reasonable, the EPA many years ago specified a set of work practices to keep emissions to a minimum—for example, that asbestos should be wetted down before handling and disposed of carefully. The building industry called for more specifics. Modifications in the Clean Air Act eventually permitted

---

[13] Robert W. Kates, ed., *Managing Technological Hazards: Research Needs and Opportunities* (Boulder, CO: Institute of Behavioral Science, University of Colorado, 1977), 32.

[14] William W. Lowrance, *Of Acceptable Risk* (Los Altos, CA: William Kaufmann, 1976), 134.

the EPA to issue enforceable rules on work practices, and now the Occupational Safety and Health Administration is also involved.

Society's attempts at regulation have indeed often failed, but it would be wrong to write off rule-making and rule-following as futile. Good laws, effectively enforced, clearly produce benefits. They authoritatively establish reasonable minimal standards of professional conduct and provide at least a self-interested motive for most people and corporations to comply. Moreover, they serve as a powerful support and defense for those who wish to act ethically in situations where ethical conduct might be less than welcome.

Engineering as social experimentation can provide engineers with a proper perspective on laws and regulations in that rules that govern engineering practice should not be devised or construed as rules of a game but as rules of responsible experimentation. Such a view places proper responsibility on the engineer who is intimately connected with his or her "experiment" and responsible for its safe conduct.

Moreover, it suggests the following conclusions: Precise rules and enforceable sanctions are appropriate in cases of ethical misconduct that involve violations of well-established and regularly reexamined engineering procedures that have as their purpose the safety and well-being of the public. Little of an experimental nature is probably occurring in such standard activities, and the type of professional conduct required is most likely very clear. In areas where experimentation is involved more substantially, however, rules must not attempt to cover all possible outcomes of an experiment, nor must they force engineers to adopt rigidly specified courses of action. It is here that regulations should be broad, but written to hold engineers accountable for their decisions. Through their professional societies engineers should also play an active role in establishing (or changing) enforceable rules as well as in enforcing them.

Industrial Standards

There is one area in which industry usually welcomes greater specificity, and that is in regard to standards. Product standards facilitate the interchange of components, they serve as ready-made substitutes for lengthy design specifications, and they decrease production costs.

Standards consist of explicit specifications that, when followed with care, ensure that stated criteria for interchangeability and quality will be attained. Examples range from automobile tire sizes and load ratings to computer protocols. Table 4–1 lists purposes of standards and gives some examples to illustrate those purposes.

Table 4–1 Types of standards

Criterion	Purpose	Selected examples
Uniformity of physical properties and functions	Accuracy in measurement, interchangeability, ease of handling	Standards of weights, screw dimensions, standard time, film size
Safety and reliability	Preparation of injury, death, and loss of income or property	National Electric Code, boiler code, methods of handling toxic wastes
Quality of product	Fair value for price	Plywood grade, lamp life
Quality of personnel and service	Competence in carrying out tasks	Accreditation of schools, professional licenses
Use of accepted procedures	Sound design, ease of communications	Drawing symbols, test procedures
Separability	Freedom from interference	Highway lane markings, radio frequency bands
Quality procedures approved by ISO	Assurance of product acceptance in member countries	Quality of products, work, certificates, and degrees

ISO, International Organization for Standardization

Standards are established by companies for in-house use and by professional associations and trade associations for industry-wide use. They may also be prescribed as parts of laws and official regulations, for example, in mandatory standards, which often arise from lack of adherence to voluntary standards.

Standards not only help the manufacturers, they also benefit the client and the public. They preserve some competitiveness in industry by reducing overemphasis on name brands and giving the smaller manufacturer a chance to compete. They ensure a measure of quality and thus facilitate more realistic trade-off decisions. International standards are becoming a necessity in European and world trade. An interesting approach has been adopted by the International Standards Organization (ISO) that replaces the detailed national specifications for a plethora of products with statements of procedures that a manufacturer guarantees to carry out to assure quality products.

Standards have been a hindrance at times. For many years they were mostly descriptive, specifying, for instance, how many joists of what size should support a given type of floor. Clearly such standards tended to stifle innovation. The move to performance standards, which in the case of a floor may specify only the required load-bearing capacity, has alleviated that problem somewhat. But other difficulties can arise when special interests (e.g., manufacturers, trade unions, exporters, and importers) manage to impose unnecessary provisions on standards or remove important provisions from them to secure their own narrow self-interest. Requiring metal conduits for home wiring is one example of this problem. Modern conductor coverings have eliminated the need for metal conduit in many applications, but many localities still require it. Its use sells more conduit and labor time for installation. But standards did not foresee the dangers encountered when aluminum was substituted for copper as conductor in home wiring, as happened in the United States during the copper scarcity occasioned by the Vietnam War. Until better ways were devised for fastening aluminum conductors, many fires occurred because of the gradual loosening of screw terminals.

Nevertheless, there are standards nowadays for practically everything, it seems, and consequently we often assume that stricter regulation exists than may actually be the case. The public tends to trust the National Electrical Code in all matters of power distribution and wiring, but how many people realize that this code, issued by the National Fire Protection Association, is primarily oriented toward fire hazards? Only recently have its provisions against electric shock begun to be strengthened. Few consumers know that an Underwriter Laboratories seal prominently affixed to the cord of an electrical appliance may pertain only to the cord and not to the rest of the device. In a similar vein, a patent notation inscribed on the handle of a product may refer just to the handle, and then possibly only to the design of the handle's appearance.

Challenger

Let us apply this discussion of engineering as social experimentation to the explosion of the space shuttle *Challenger*, and by extension the space shuttle *Columbia*. The *Columbia* and its sister ships, *Challenger, Discovery,* and *Endeavor,* were delta-wing craft with a huge payload bay (Figure 4-1). Early, sleek designs had to be abandoned to satisfy U.S. Air Force requirements when the Air Force was ordered to use the National Aeronautics and Space Administration (NASA) shuttle instead of its own expendable rockets for launching satellites and other missions. As shown in Figure 4–2, each orbiter has three main engines fueled by several million pounds of liquid hydrogen; the fuel is carried in an immense, external, divided fuel tank, which is jettisoned when empty. During liftoff the main engines fire for approximately 8.5 minutes, although during the first 2 minutes of the launch much of the thrust is provided by two booster rockets. These are of the solid-fuel type, each burning a one-million-pound load of a mixture of aluminum, potassium chloride, and iron oxide.

The casing of each booster rocket is approximately 150 feet long and 12 feet in diameter. It consists of cylindrical segments

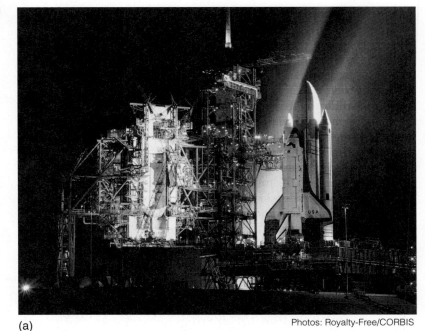

Figure 4–1
(a) Space shuttle
Columbia.
(b) Space shuttle
Challenger.
(c) Space shuttle
Discovery.
(d) Space shuttle
Endeavour.

(a)

Photos: Royalty-Free/CORBIS

(b)

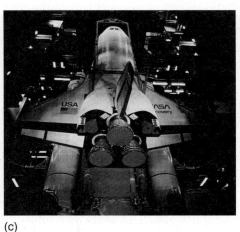

(c)

(d)

that are assembled at the launch site. The four field joints use seals composed of pairs of O-rings made of vulcanized rubber. The O-rings work in conjunction with a putty barrier of zinc chromide.

The shuttle flights were successful, although not as frequent as had been hoped. NASA tried hard to portray the shuttle program as an operational system that could pay for itself. But aerospace engineers intimately involved in designing, manufacturing,

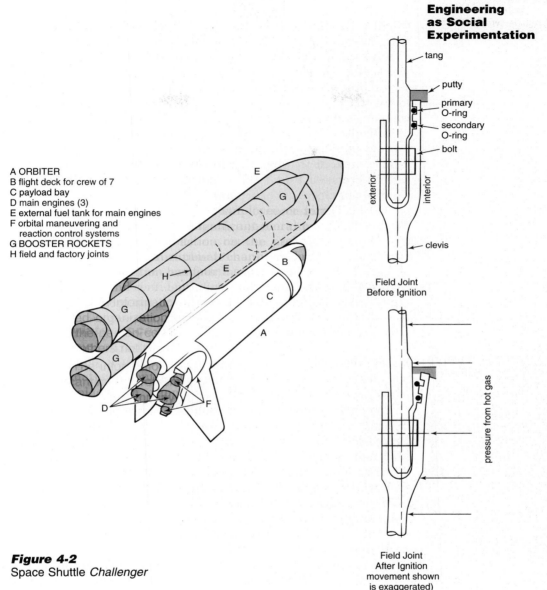

A ORBITER
B flight deck for crew of 7
C payload bay
D main engines (3)
E external fuel tank for main engines
F orbital maneuvering and
   reaction control systems
G BOOSTER ROCKETS
H field and factory joints

tang

putty

primary
O-ring

secondary
O-ring

bolt

exterior

interior

clevis

Field Joint
Before Ignition

pressure from hot gas

Field Joint
After Ignition
movement shown
is exaggerated)

Figure 4-2
Space Shuttle *Challenger*

assembling, testing, and operating the shuttle still regarded it
as an experimental undertaking in 1986. These engineers were
employees of manufacturers, such as Rockwell International
(orbiter and main rocket) and Morton-Thiokol (booster rockets),
or they worked for NASA at one of its several centers: Marshall
Space Flight Center, Huntsville, Alabama (responsible for the
propulsion system); Kennedy Space Center, Cape Kennedy, Florida (launch operations); Johnson Space Center, Houston, Texas

(flight control); and the office of the chief engineer, Washington, D.C. (overall responsibility for safety, among other duties).

After embarrassing delays, *Challenger's* first flight for 1986 was set for Tuesday morning, January 28. But Allan J. McDonald, who represented Morton-Thiokol at Cape Kennedy, was worried about the freezing temperatures predicted for the night. As his company's director of the solid-rocket booster project, he knew of difficulties that had been experienced with the field joints on a previous cold-weather launch when the temperature had been mild compared to what was forecast. He therefore arranged a teleconference so that NASA engineers could confer with Morton-Thiokol engineers at their plant in Utah.

Arnold Thompson and Roger Boisjoly, two seal experts at Morton-Thiokol, explained to their own colleagues and managers as well as the NASA representatives how on launch the booster rocket walls bulge, and the combustion gases can blow past one or even both of the O-rings that make up the field joints (see Figure 4–2).[15] The rings char and erode, as had been observed on many previous flights. In cold weather the problem is aggravated because the rings and the putty packing are less pliable then. But only limited consideration was given to the past history of O-ring damage in terms of temperature. Consideration of the entire launch temperature history indicates that the probability of O-ring distress is increased to almost a certainty if the temperature of the joint is less than 65°F.[16]

The engineering managers, Bob Lund (vice president of engineering) and Joe Kilminster (vice president for booster rockets), agreed that there was a problem with safety. The team from Marshall Space Flight Center was incredulous. Because the specifications called for an operating temperature of the solid fuel prior to combustion of 40°F to 90°F, one could surely allow lower or higher outdoor temperatures, notwithstanding Boisjoly's testimony and recommendation that no launch should occur at less than 53°F. They were clearly annoyed at facing yet another postponement.

Top executives of Morton-Thiokol were also sitting in on the teleconference. Their concern was the image of the company, which was in the process of negotiating a renewal of the booster rocket contract with NASA. During a recess Senior Vice President Jerry Mason turned to Bob Lund and told him "to take off your

---

[15] Wade Robison et al., "Representation and Misrepresentation: Tufte and the Morton-Thiokol Engineers on the *Challenger*," *Science and Engineering Ethics* 8, no. 1 (2002): 59–81.

[16] Rogers Commission Report, *Report of the Presidential Commission on the Space Shuttle Challenger Accident* (Washington, DC: U.S. Government Printing Office, 1986).

engineering hat and put on your management hat." It was a subsequent vote (of the managers only) that produced the company's official finding that the seals could not be shown to be unsafe. The engineers' judgment was not considered sufficiently weighty. At Cape Kennedy, Allan McDonald refused to sign the formal recommendation to launch; Joe Kilminster had to. Accounts of the *Challenger* disaster tell of the cold Tuesday morning, the high seas that forced the recovery ships to seek coastal shelter, the ice at the launch site, and the concern expressed by Rockwell engineers that the ice might shatter and hit the orbiter or rocket casings.[17] The inability of these engineers to *prove* that the liftoff would be unsafe was taken by NASA as an approval by Rockwell to launch.

The countdown ended at 11:38 AM. The temperature had risen to 36°F. As the rockets carrying *Challenger* rose from the ground, cameras recorded puffs of smoke that emanated from one of the field joints on the right booster rocket. Soon these turned into a flame that hit the external fuel tank and a strut holding the booster rocket. The hydrogen in the tank caught fire, the booster rocket broke loose, smashed into *Challenger's* wing, then into the external fuel tank. At 76 seconds into the flight, by the time *Challenger* and its rockets had reached 50,000 feet, it was totally engulfed in a fireball. The crew cabin separated and fell into the ocean, killing all aboard: Mission Commander Francis (Dick) Scobee; Pilot Michael Smith; Mission Specialists Gregory Jarvis, Ronald McNair, Ellison Onizuka, Judith Resnik; and "teacher in space" Christa MacAuliffe.

Why was safe operation of the space shuttle not stressed more? First of all, we must remember that the shuttle program was indeed still a truly experimental and research undertaking. Next, it is quite clear that the members of the crews knew that they were embarking on dangerous missions. But it has also been revealed that the *Challenger* astronauts were not informed of particular problems such as the field joints. They were not asked for their consent to be launched under circumstances that experienced engineers had claimed to be unsafe and without any safe escape mechanism (safe exit) available should things go wrong.

The reason for the rather cavalier attitude toward safety is revealed in the way NASA assessed the system's reliability. For instance, recovered booster rocket casings had indicated that the

---

[17] Malcolm McConnell, *Challenger, a Major Malfunction* (Garden City, NY: Doubleday, 1987); Rosa Lynn B. Pinkus et al., *Engineering Ethics: Balancing Cost, Schedule, and Risk—Lessons Learned from the Space Shuttle* (Cambridge: Cambridge University Press, 1997); Diane Vaughan, *The Challenger Launch Decision: Risky Technology, Culture, and Deviance at NASA* (Chicago: University of Chicago Press, 1996).

field-joint seals had been damaged in many of the earlier flights. The waivers necessary to proceed with launches had become mere gestures. Richard Feynman made the following observations as a member of the Presidential Commission on the Space Shuttle Challenger Accident (called the Rogers Commission after its chairman): "I read all of these [NASA flight readiness] reviews and they agonize whether they can go even though they had some blow-by in the seal or they had a cracked blade in the pump of one of the engines . . . and they decide 'yes.' Then it flies and nothing happens. Then it is suggested . . . that the risk is no longer so high. For the next flight we can lower our standards a little bit because we got away with it last time . . . It is a kind of Russian roulette."[18]

Since the early days of unmanned space flight, approximately 1 in every 25 solid-fuel rocket boosters failed. Given improvements over the years, Feynman thought that 1 in every 50 to 100 might be a reasonable estimate now. Yet NASA counted on only one crash in every 100,000 launches.

Another area of concern was NASA's unwillingness to wait out risky weather. When serving as weather observer, astronaut John Young was dismayed to find his recommendations to postpone launches disregarded several times. Things had not changed much by March 26, 1987, when NASA ignored its devices monitoring electric storm conditions, launched a Navy communications satellite atop an Atlas-Centaur rocket, and had to destroy the $160 million system when it veered off course after being hit by lightning. The monitors had been installed after a similar event involving an Apollo command module eighteen years before had nearly aborted a trip to the moon.

Veteran astronauts were also dismayed at NASA management's decision to land at Cape Kennedy as often as possible despite its unfavorable landing conditions, including strong crosswinds and changeable weather. The alternative, Edwards Air Force Base in California, is a better landing place but necessitates a piggyback ride for the shuttle on a Boeing 747 home to Florida. This costs time and money.

In 1982 Albert Flores had conducted a study of safety concerns at the Johnson Space Center. He found its engineers to be strongly committed to safety in all aspects of design. When they were asked how managers might further improve safety awareness, there were few concrete suggestions but many comments on how safety concerns were ignored or negatively affected by management. One engineer was quoted as saying, "A small amount of professional safety effort and upper management support can

---

[18] *Rogers Commission Report,* Report of the Presidential Commission.

cause a quantum safety improvement with little expense."[19] This points to the important role of management in building a strong sense of responsibility for safety first and schedules second.

The space shuttle's field joints are designated criticality 1, which means there is no backup. Therefore a leaky field joint will result in failure of the mission and loss of life. There are 700 items of criticality 1 on the shuttle. A problem with any one of them should have been cause enough to do more than just launch more shuttles without modification while working on a better system. Improved seal designs had already been developed, but the new rockets would not have been ready for some time. In the meantime, the old booster rockets should have been recalled.

In several respects the ethical issues in the *Challenger* case resemble those of other such cases. Concern for safety gave way to institutional posturing. Danger signals did not go beyond Morton-Thiokol and Marshall Space Flight Center in the *Challenger* case. No effective recall was instituted. There were concerned engineers who spoke out, but ultimately they felt it only proper to submit to management decisions.

One notable aspect of the *Challenger* case is the late-hour teleconference that Allan McDonald had arranged from the *Challenger* launch site to get knowledgeable engineers to discuss the seal problem from a technical viewpoint. This tense conference did not involve lengthy discussions of ethics, but it revealed the virtues (or lack thereof) that allow us to distinguish between the "right stuff" and the "wrong stuff." This is well described by one aerospace engineer as arrogance, specifically, "The arrogance that prompts higher-level decision makers to pretend that factors other than engineering judgment should influence flight safety decisions and, more important, the arrogance that rationalizes overruling the engineering judgment of engineers close to the problem by those whose expertise is naive and superficial by comparison."[20] Included, surely, is the arrogance of those who reversed NASA's (paraphrased) motto "Don't fly if it cannot be shown to be safe" to "Fly unless it can be shown not to be safe."

In a speech to engineering students at the Massachusetts Institute of Technology a year after the *Challenger* disaster, Roger Boisjoly said: "I have been asked by some if I would testify again if I knew in advance of the potential consequences to me and my career. My answer is always an immediate yes. I couldn't live with any self-respect if I tailored my actions based

---

[19] Albert Flores, ed., *Designing for Safety: Engineering Ethics in Organizational Contexts* (Troy, NY: Rensselaer Polytechnic Institute, 1982), 79.
[20] Calvin E. Moeller, "*Challenger* Catastrophe," *Los Angeles Times*, letters to the editor, March 11, 1986.

on potential personal consequences as a result of my honorable actions."[21]

Discussion Questions

1. A common excuse for carrying out a morally questionable project is, "If I don't do it somebody else will." This rationale may be tempting for engineers who typically work in situations where someone else might be ready to replace them on a project. Do you view it as a legitimate excuse for engaging in projects that might be unethical? In your answer, comment on the concept of responsible conduct developed in this section.

2. Another commonly used phrase, "I only work here," implies that one is not personally accountable for the company rules because one does not make them. It also suggests that one wishes to restrict one's area of responsibility within tight bounds as defined by those rules. In light of the discussion in this section, respond to the potential implications of this phrase and the attitude it represents when exhibited by engineers.

3. Threats to a sense of personal responsibility are neither unique to, nor more acute for, engineers than they are for others involved with engineering and its results. The reason is that, in general, public accountability also tends to lessen as professional roles become narrowly differentiated. With this in mind, critique each of the remarks made in the following dialogue. Is the remark true, or partially true? What needs to be added to make it accurate?

Engineer: My responsibility is to receive directives and to create products within specifications set by others. The decision about what products to make and their general specifications are economic in nature and made by management.

Scientist: My responsibility is to gain knowledge. How the knowledge is applied is an economic decision made by management, or else a political decision made by elected representatives in government.

Manager: My responsibility is solely to make profits for stockholders.

Stockholder: I invest my money for the purpose of making a profit. It is up to our boards and managers to make decisions about the directions of technological development.

---

[21] Roger M. Boisjoly, "The Challenger Disaster: Moral Responsibility and the Working Engineers," (speech on shuttle disaster delivered to MIT students, January 7, 1987). *Books and Religion* 15 (March–April 1987): 3.

Consumer: My responsibility is to my family. Government should make sure corporations do not harm me with dangerous products, harmful side effects of technology, or dishonest claims.

Government Regulator: By current reckoning, government has strangled the economy through overregulation of business. Accordingly, at present on my job, especially given decreasing budget allotments, I must back off from the idea that business should be policed, and urge corporations to assume greater public responsibility.

4. Mismatched bumpers: What happens when a passenger car rear-ends a truck or a sports utility vehicle (SUV)? The bumpers usually ride at different heights, so even modest collisions can result in major repair bills. (At high speed, with the front of the car nose down when braking, people in convertibles have been decapitated on contact devoid of protection by bumpers.) Ought there to be a law?

5. Chairman Rogers asked Bob Lund: "Why did you change your decision [that the seals would not hold up] when you changed hats?" What might motivate you, as a midlevel manager, to go along with top management when told to "take off your engineering hat and put on your management hat"? Applying the engineering-as-experimentation model, what might responsible experimenters have done in response to the question?

6. Under what conditions would you say it is safe to launch a shuttle without an escape mechanism for the crew? Also, discuss the role of the astronauts in shuttle safety. To what extent should they (or at least the orbiter commanders) have involved themselves more actively in looking for safety defects in design or operation?

7. Examine official reports to determine to what extent the *Columbia* explosion can be ascribed to technical and/or management deficiencies. Was there a failure to learn from earlier events? (Search the Web under "CAIB Report": the Columbia Accident Investigation Board Report.)

CHAPTER 5

Commitment to Safety

Pilot Dan Gellert was flying an Eastern Airlines Lockheed L-1011, cruising at an altitude of 10,000 feet, when he inadvertently dropped his flight plan.[1] Being on autopilot control, he casually leaned down to pick it up. In doing so, he bumped the control stick. This should not have mattered, but immediately the plane went into a steep dive, terrifying the 230 passengers. Badly shaken himself, Gellert was nevertheless able to grab the control stick and ease the plane back on course. Although much altitude had been lost, the altimeter still read 10,000 feet.

Not long before this incident, one of Gellert's colleagues had been in a flight trainer when the autopilot and the flight trainer disengaged, producing a crash on an automatic landing approach. Fortunately it all happened in simulation. But just a short time later, an Eastern Airlines L-1011 actually crashed on approach to Miami. On that flight there seemed to have been some problem with the landing gear, so the plane had been placed on autopilot at 2,000 feet while the crew investigated the trouble. Four minutes later, after apparently losing altitude without warning while the crew was distracted, it crashed in the Everglades, killing 103 people.

A year later Gellert was again flying an L-1011, and the autopilot disengaged once more when it should not have done so. The plane was supposedly at 500 feet and on the proper glide slope to landing as it broke through a cloud cover. Suddenly realizing it was only at 200 feet and above a densely populated area, the crew had to engage the plane's full takeoff power to make the runway safely.

The L-1011 incidents point out how vulnerable our intricate machines and control systems can be, how they can malfunction because of unanticipated circumstances, and how important it

---

[1] Dan Gellert, "Whistle-Blower: Dan Gellert, Airline Pilot," *Civil Liberties Review* (September 1978).

is to design for proper human-machine interactions whenever human safety is involved. In this chapter we discuss the role of safety as seen by the public and the engineer.

Typically, several groups of people are involved in safety issues, each with its own interests at stake. If we now consider that within each group there are differences of opinion regarding what is safe and what is not, it becomes obvious that "safety" can be an elusive term, as can "risk." Following a look at these basic concepts, we will turn to safety and risk assessment and methods of reducing risk. Finally, in examining the nuclear power plant accident at Three Mile Island, we will consider the implications of an ever-growing complexity in engineered systems and the ultimate need for safe exits, that is, for designs and procedures ensuring that if a product fails it will fail safely, and the user can avoid harm.

5.1 Safety and Risk

We demand safe products and services, but we also realize that we may have to pay for this safety. To complicate matters, what may be safe enough for one person may not be for someone else—either because of different perceptions about what is safe or because of different predispositions to harm. A power saw in the hands of a child will never be as safe as it can be in the hands of an adult. And, a sick adult is more prone to suffer ill effects from air pollution than is a healthy adult.

Absolute safety, in the senses of (a) entirely risk-free activities and products, or (b) a degree of safety that satisfies all individuals or groups under all conditions, is neither attainable nor affordable. Yet it is important that we come to some understanding of what we mean by safety.

The Concept of Safety

One approach to defining safety would be to render the notion thoroughly subjective by defining it in terms of whatever risks a person judges to be acceptable. Such a definition was given by William W. Lowrance: "A thing is safe if its risks are judged to be acceptable."[2] This approach helps underscore the notion that judgments about safety are tacitly value judgments about what is acceptable risk to a given person or group. Differences in appraisals of safety are thus correctly seen as reflecting differences in values.

Lowrance's definition, however, needs to be modified, for it departs too far from our common understanding of safety. This can be shown if we consider three types of situations. Imagine,

---

[2] William W. Lowrance, *Of Acceptable Risk* (Los Altos, CA: William Kaufmann, 1976), 8.

first, a case where we seriously underestimate the risks of something, say of using a toaster we see at a garage sale. On the basis of that mistaken view, we judge it to be very safe and buy it. On taking it home and trying to make toast with it, however, it sends us to the hospital with a severe electric shock or burn. Using the ordinary notion of safety, we conclude we were wrong in our earlier judgment: The toaster was not safe at all. Given our values and our needs, its risks should not have been judged acceptable earlier. Yet, by Lowrance's definition, we would be forced to say that prior to the accident the toaster was entirely safe because, after all, at that time we had judged the risks to be acceptable.

Consider, second, the case where we grossly overestimate the risks of something. For example, we irrationally think fluoride in drinking water will kill a fifth of the populace. According to Lowrance's definition, the fluoridated water is unsafe, because we judge its risks to be unacceptable.

Third, there is the situation in which a group makes no judgment at all about whether the risks of a thing are acceptable or not—they simply do not think about it. By Lowrance's definition, this means the thing is neither safe nor unsafe with respect to that group. Yet this goes against our ordinary ways of thinking about safety. For example, we normally say that some cars are safe and others unsafe, even though many people may never even think about the safety of the cars they drive.

There must be at least some objective point of reference outside ourselves that allows us to decide whether our judgments about safety are correct once we have settled on what constitutes to us an acceptable risk. An expanded definition could capture this element, without omitting the insight already noted that safety judgments are relative to people's value perspectives.[3] One option is simply to equate safety with the absence of risk. Because little in life, and nothing in engineering, is risk-free, we prefer to adopt a modified version of Lowrance's definition: *A thing is safe if, were its risks fully known, those risks would be judged acceptable by reasonable persons in light of their settled value principles.*

In our view, then, safety is a matter of how people would find risks acceptable or unacceptable if they knew the risks and were basing their judgments on their most settled value perspectives. To this extent safety is an *objective* matter. It is a *subjective* matter to the extent that value perspectives differ. In what follows we will usually speak of safety simply as acceptable risk. But this is

---

[3] Council for Science and Society, *The Acceptability of Risks* (England: Barry Rose, Ringwood, Hants, 1977), 3.

merely for convenience, and it should be interpreted as an endorsement of Lowrance's definition only as we have qualified it.

Safety is often thought of in terms of degrees and comparisons. We speak of something as "fairly safe" or "relatively safe" (compared with similar things). Using our definition, this translates as the degree to which a person or group, judging on the basis of their settled values, would decide that the risks of something are more or less acceptable in comparison with the risks of some other thing, and in light of relevant information. For example, when we say that airplane travel is safer than automobile travel, we mean that for each mile traveled it leads to fewer deaths and injuries—the risky elements that our settled values lead us to avoid. Finally, we interpret "things" to include products, services, institutional processes, and disaster protection.

Risks

We say a thing is not safe if it exposes us to unacceptable risk; but what is meant by "risk"? *A risk is the potential that something unwanted and harmful may occur.* We take a risk when we undertake something or use a product or substance that is not safe. William D. Rowe refers to the "potential for the realization of unwanted consequences from impending events."[4] Thus a future, possible occurrence of harm is postulated.

Risk, like harm, is a broad concept covering many different types of unwanted occurrences. In regard to technology, it can equally well include dangers of bodily harm, of economic loss, or of environmental degradation. These in turn can be caused by delayed job completion, faulty products or systems, or economically or environmentally injurious solutions to technological problems.

Good engineering practice has always been concerned with safety. But as technology's influence on society has grown, so has public concern about technological risks increased. In addition to measurable and identifiable hazards arising from the use of consumer products and from production processes in factories, some of the less obvious effects of technology are now also making their way to public consciousness. Although the latter are often referred to as new risks, many of them have existed for some time. They are new only in the sense that (1) they are now identifiable—because of changes in the magnitude of the risks they present, because they have passed a certain threshold of accumulation in our environment, or because of a change in measuring techniques, or (2) the public's perception of them has

---

[4] William D. Rowe, *An Anatomy of Risk* (New York: John Wiley & Sons, 1977), 24.

changed—because of education, experience, media attention, or a reduction in other hitherto dominant and masking risks.

Meanwhile, natural hazards continue to threaten human populations. Technology has greatly reduced the scope of some of these, such as floods, but at the same time it has increased our vulnerability to other natural hazards, such as earthquakes, as they affect our ever-greater concentrations of population and cause greater damage to our finely tuned technological networks of long lifelines for water, energy, and food. Of equal concern are our disposal services (sewers, landfills, recovery and neutralizing of toxic wastes) and public notification of potential hazards they present.

Acceptability of Risk

Having adopted a modified version of Lowrance's definition of safety as acceptable risk, we need to examine the idea of acceptability more closely. William D. Rowe says that "a risk is acceptable when those affected are generally no longer (or not) apprehensive about it."[5] Apprehensiveness depends to a large extent on how the risk is perceived. This is influenced by such factors as (1) whether the risk is accepted voluntarily; (2) the effects of knowledge on how the probabilities of harm (or benefit) are known or perceived; (3) if the risks are job-related or other pressures exist that cause people to be aware of or to overlook risks; (4) whether the effects of a risky activity or situation are immediately noticeable or are close at hand; (5) and whether the potential victims are identifiable beforehand. Let us illustrate these elements of risk perception.

Voluntarism and Control. John and Ann Smith and their children enjoy riding motorcycles over rough terrain for amusement. They take voluntary risks, part of being engaged in such a potentially dangerous sport. They do not expect the manufacturer of their dirt bikes to adhere to the same standards of safety as they would the makers of a passenger car used for daily commuting. The bikes should be sturdy, but guards covering exposed parts of the engine, padded instrument panels, collapsible steering mechanisms, or emergency brakes are clearly unnecessary, if not inappropriate.

In discussing dirt bikes and the like we do not include all-terrain three-wheel vehicles. Those represent hazards of greater magnitude because of the false sense of security they give the rider. They tip over easily. During the five years before they were

---

[5] William D. Rowe, "What Is an Acceptable Risk and How Can It Be Determined?" in *Energy Risk Management*, ed. G. T. Goodman and W. D. Rowe (New York: Academic Press, 1979), 328.

forbidden in the United States, they were responsible for nearly 900 deaths and 300,000 injuries. Approximately half of the casualties were children younger than 16 years old.

John and Ann live near a chemical plant. It is the only area in which they can afford to live, and it is near the shipyard where they both work. At home they suffer from some air pollution, and there are some toxic wastes in the ground. Official inspectors tell them not to worry. Nevertheless they do, and they think they have reason to complain—they do not care to be exposed to risks from a chemical plant with which they have no relationship except on an involuntary basis. Any beneficial link to the plant through consumer products or other possible connections is very remote and, moreover, subject to choice.

John and Ann behave as most of us would under the circumstances: We are much less apprehensive about the risks to which we expose ourselves voluntarily than about those to which we are exposed involuntarily. In terms of our "engineering as social experimentation" paradigm, people are more willing to be the subjects of their own experiments (social or not) than of someone else's.

Intimately connected with this notion of voluntarism is the matter of control. The Smiths choose where and when they will ride their bikes. They have selected their machines, and they are proud of how well they can control them, or think they can. They are aware of accident figures, but they tell themselves those apply to other riders, not to them. In this manner they may well display the characteristically unrealistic confidence of most people when they believe hazards to be under their control.[6] But still, riding motorbikes, skiing, hang gliding, bungee jumping, horseback riding, boxing, and other hazardous sports are usually carried out under the assumed control of the participants. Enthusiasts worry less about their risks than the dangers of, say, air pollution or airline safety. Another reason for not worrying so much about the consequences of these sports is that rarely does any one accident injure innocent bystanders.

Effect of Information on Risk Assessments. The manner in which information necessary for decision making is presented can greatly influence how risks are perceived. The Smiths are careless

---

[6] Paul Slovic, Baruch Fischhoff, and Sarah Lichtenstein, "Weighing the Risks: Which Rights Are Acceptable?" *Environment 21* (April 1979): 14–20 and (May 1979): 17–20, 32–38; Paul Slovic, Baruch Fischhoff, and Sarah Lichtenstein, "Risky Assumptions," *Psychology Today* 14 (June 1980): 44–48.

about using seat belts in their car. They know that the probability of their having an accident on any one trip is small. Had they been told, however, that in the course of 50 years of driving, at 800 trips per year, there is a probability of 1 in 3 that they will receive at least one disabling injury, then their seat-belt habits, and their attitude about seat-belt laws, would likely be different.[7]

Studies have verified that a change in the manner in which information about a danger is presented can lead to a striking reversal of preferences about how to deal with that danger. Consider, for example, an experiment in which two groups of 150 people were told about the strategies available for combating a disease (that in some ways foreshadowed the severe acute respiratory syndrome [SARS] epidemic in 2003). The first group was given the following description:

> Imagine that the U.S. is preparing for the outbreak of an unusual Asian disease, which is expected to kill 600 people. Two alternative programs to combat the disease have been proposed. Assume that the exact scientific estimate of the consequences of the programs are as follows: If Program A is adopted, 200 people will be saved. If Program B is adopted, there is 1/3 probability that 600 people will be saved, and 2/3 probability that no people will be saved. Which of the two programs would you favor?[8]

The researchers reported that 72 percent of the respondents selected program A, and only 28 percent selected program B. Evidently the vivid prospect of saving 200 people led many of them to feel averse to taking a risk on possibly saving all 600 lives.

The second group was given the same problem and the same two options, but the options were worded differently: "If Program C is adopted, 400 people will die. If Program D is adopted, there is 1/3 probability that nobody will die and 2/3 probability that 600 people will die. Which of the two programs would you favor?"

This time only 22 percent chose program C, which is the same as program A. Seventy-eight percent chose program D, which is identical to program B.

One conclusion that we draw from the experiment is that options perceived as yielding firm gains will tend to be preferred over those from which gains are perceived as risky or only probable. A second conclusion is that options emphasizing firm losses will tend to be avoided in favor of those with chances of success

[7] Richard J. Arnould and Henry Grabowski, "Auto Safety Regulation: An Analysis of Market Failure," *Bell Journal of Economics* 12 (Spring 1981): 35.

[8] Amos Tversky and Daniel Kahneman, "The Framing of Decisions and the Psychology of Choice," *Science* 211 (January 30, 1981): 453.

that are perceived as probable. In short, people tend to be more willing to take risks to avoid perceived firm losses than they are to win only possible gains.

Job-Related Risks. John Smith's work in the shipyard has in the past exposed him to asbestos. He is aware now of the high percentage of asbestosis cases among his coworkers, and after consulting his own physician finds that he is slightly affected himself. Even Ann, who works in a clerical position at the shipyard, has shown symptoms of asbestosis as a result of handling her husband's clothes. Earlier John saw no point to "all the fuss stirred up by some do-gooders." He figured that he was being paid to do a job; he felt the masks that were occasionally handed out gave him sufficient protection, and he thought the company physician was giving him a clean bill of health.

In this regard, John's thinking is similar to that of many workers who take risks on their jobs in stride, and sometimes even approach them with bravado. Of course, exposure to risks on a job is in a sense voluntary since one can always refuse to submit oneself to them, and workers perhaps even have some control over how their work is carried out. But often employees have little choice other than to stick with what is for them the only available job and to do as they are told. What they are often not told about is their exposure to toxic substances and other dangers that cannot readily be seen, smelled, heard, or otherwise sensed.

Unions and occupational health and safety regulations (such as right-to-know rules regarding toxics) can correct the worst situations, but standards regulating conditions in the workplace (its air quality, for instance) are generally still far below those that regulate conditions in our general (public) environment. It may be argued that the "public" encompasses many people of only marginal health whose low thresholds for pollution demand a fairly clean environment. Conversely, factory workers are seldom carefully screened for their work. And in all but the most severe environments (those conducive to black lung or brown lung, for instance), unions display little desire for change, lest necessary modifications of the workplace force employers out of business.

Magnitude and Proximity. Our reaction to risk is affected by the dread of a possible mishap, both in terms of its magnitude and of the personal identification or relationship we may have with the potential victims. A single major airplane crash in a remote country, the specter of a child we know or observe on the television screen trapped in a cave-in—these affect us more acutely than the ongoing but anonymous carnage on the highways, at least until someone close to us is involved in a car accident.

In terms of numbers alone we feel much more keenly about a potential risk if one of us out of a group of 20 intimate friends is likely to be subjected to great harm than if it might affect, say, 50 strangers out of a proportionally larger group of 1,000. This proximity effect arises in perceptions of risk over time as well. A future risk is easily dismissed by various rationalizations including (1) the attitude of "out of sight, out of mind," (2) the assumption that predictions for the future must be discounted by using lower probabilities, or (3) the belief that a countermeasure will be found in time.

Engineers face two problems with public conceptions of safety. On the one hand, there is the overly optimistic attitude that things that are familiar, that have not hurt us before, and over which we have some control, present no real risks. On the other hand, there is the dread people feel when an accident kills or maims in large numbers, or harms those we know, even though statistically speaking such accidents might occur infrequently.

Discussion Questions

1. Describe a real or imagined traffic problem in your neighborhood involving children and elderly people who find it difficult to cross a busy street. Put yourself in the position of (a) a commuter traveling to work on that street; (b) the parent of a child, or the relative of an older person who has to cross that street on occasion; (c) a police officer assigned to keep the traffic moving on that street; and (d) the town's traffic engineer working under a tight budget.

   Describe how in these various roles you might react to (e) complaints about conditions dangerous to pedestrians at that crossing and (f) requests for a pedestrian crossing protected by traffic or warning lights.

2. In some technologically advanced nations, a number of industries that have found themselves restricted by safety regulations have resorted to dumping their products on—or moving their production processes to—less-developed countries where higher risks are tolerated. Examples are the dumping of unsafe or ineffective drugs on Third World countries by pharmaceutical companies from highly industrialized countries, and in the past the transfer of asbestos processing from the United States to Mexico.[9]

---

[9] Milton Silverman, Philip Lee, and Mia Lydecker, *Prescription for Death: The Drugging of the Third World* (San Francisco: University of California Press, 1981); and Henry Shue, "Exporting Hazards," *Ethics* 91 (July 1981): 586.

More recently, toxic wastes—from lead-acid batteries to nuclear wastes—have been added to the list of "exports." To what extent do differences in perception of risk justify the transfer of such merchandise and production processes to other countries? Is this an activity that can or should be regulated?

3. Grain dust is pound for pound more explosive than coal dust or gunpowder. Ignited by an electrostatic discharge or other cause, it has ripped apart grain silos and has killed or wounded many workers over the years. When 54 people were killed during Christmas week 1977, grain handlers and the U.S. government finally decided to combat dust accumulation.[10] Ten years, 59 deaths, and 317 serious injuries later, a compromise standard was agreed on that designates dust accumulation of one-eighth inch or more as dangerous and impermissible in silos in the United States. Nevertheless, on Monday, June 8, 1998, a series of explosions killed seven workers performing routine maintenance at one of the largest grain elevators in the world, demolishing one of the 246 concrete, 120-feet-high silos that stretch over a length of one-half mile in Haysville, Kansas. Discuss grain facility explosions as a case study of workplace safety and rule making: Which ideas concerning acceptable risk apply here?

5.2 Assessing and Reducing Risk

Any improvement in safety as it relates to an engineered product is often accompanied by an increase in the cost of that product. Conversely, products that are not safe incur secondary costs to the manufacturer beyond the primary (production) costs that must also be taken into account—costs associated with warranty expenses, loss of customer goodwill and even loss of customers because of injuries sustained from the use of the product, litigation, possible downtime in the manufacturing process, and so forth (Figure 5–1). It is therefore important for manufacturers and users alike to reach some understanding of the risks connected with any given product and know what it might cost to reduce those risks (or not reduce them).

Uncertainties in Design

One would think that experience and historical data would provide good information about the safety of standard products. Much has been collected and published. Gaps remain, however, because (1) there are some industries where information is not freely shared—for instance, when the cost of failure is less than the cost of fixing the problem, (2) problems and their causes are often not revealed after a legal settlement has been reached

---

[10] Eliot Marshall, "Deadlock Over Explosive Dust," *Science* 222 (November 4, 1983): 485–87; discussion, 1183.

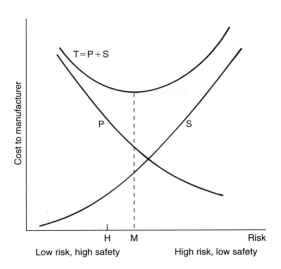

Figure 5–1
Why both low-risk and high-risk products are costly. P = primary cost of
product, including cost of safety measures involved; S = secondary costs,
including warranties, loss of customer goodwill, litigation costs, costs of
downtime, and other secondary costs. T = total cost. Minimum total cost
occurs at M, where incremental savings in primary cost (slope of P) are
offset by an equal incremental increase in secondary cost (slope of S).
Highest acceptable risk (H) may fall below risk at least cost (M), in which
case H and its higher cost must be selected as the design or operating
point.

with a condition of nondisclosure, and (3) there are always new
applications of old technology, or substitutions of materials and
components, that render the available information less useful.

Risk is seldom intentionally designed into a product. It arises
because of the many uncertainties faced by the design engineer,
the manufacturing engineer, and even the sales and applications
engineer.

To start with, there is the purpose of a design. Consider an
airliner. Is it meant to maximize profits for the airline, or is it
intended to give the highest possible return on investment? The
answer to that question is important to the company because on
it hinge a number of decisions and their outcomes and the possibility of the airline's economic success or ruin. Investing $100
million in a jet to bring in maximum profits of, say, $20 million
during a given time involves a lower return on investment than
spending $48 million on a smaller jet to bring in a return of $12
million in that same period.

Regarding applications, designs that do quite well under static
loads can fail under dynamic loading. An historical example is
the wooden bridge that collapsed when a contingent of Napoleon's

army crossed it marching in step. Such vibrations even affected one of Robert Stephenson's steel bridges, which shook violently under a contingent of marching British troops. Ever since then, soldiers are under orders to fall out of step when crossing a bridge. Wind can also cause destructive vibrations. An example is "Galloping Gertie," the Tacoma Narrows Bridge that collapsed in 1940.[11]

Apart from uncertainties about the applications of a product, there are uncertainties regarding the materials of which it is made and the level of skill that goes into designing and manufacturing it. For example, changing economic realities or hitherto unfamiliar environmental conditions such as extremely low temperatures may affect how a product is to be designed. A typical "handbook engineer" who extrapolates tabulated values without regard to their implied limits under different conditions will not fare well under such circumstances.

Caution is required even with standard materials specified for normal use. In 1981, a new bridge that had just replaced an old and trusted ferry service across the Mississippi at Prairie du Chien, Wisconsin, had to be closed because 11 of the 16 flange sections in both tie girders were found to have been fabricated from excessively brittle steel.[12] Although strength tests are (supposedly) routinely carried out on concrete, the strength of steel is all too often taken for granted.

Such drastic variations from the standard quality of a given grade of steel are exceptional; more typically the variations are small. Nevertheless the design engineer should realize that the supplier's data on items such as steel, resistors, insulation, optical glass, and so forth apply to statistical averages only. Individual components can vary considerably from the mean.

Engineers traditionally have coped with such uncertainties about materials or components, as well as incomplete knowledge about the actual operating conditions of their products, by introducing a comfortable "factor of safety." That factor is intended to protect against problems that arise when the stresses caused by anticipated loads (duty) and the stresses the product as designed is supposed to withstand (strength or capability) depart from their expected values. Stresses can be of a mechanical or other

---

[11] Henry Petroski, *To Engineer Is Human: The Role of Failure in Successful Design* (New York: St. Martin's Press, 1985); M. Levy and M. Salvadori, *Why Buildings Fall* (New York: C.C. Norton & Co., 1992); Frederic D. Schwarz, "Why Theories Fall Down," *American Heritage of Invention & Technology* (Winter 1993): 6–7.

[12] "Suit Claims Faulty Bridge Steel," *ENR* [Engineering News Record], (March 12, 1981): 14; see also (March 26, 1981): 20; (April 23, 1981): 15–16; (November 19, 1981): 28.

nature—for example, an electric field gradient to which an insulator is exposed, or the traffic density at an intersection.

A product may be said to be safe if its capability exceeds its duty. But this presupposes exact knowledge of actual capability and actual duty. In reality, the stress calculated by the engineer for a given condition of loading and the stress that ultimately materializes at that loading may vary quite a bit. This is because each component in an assembly has been allowed certain tolerances in its physical dimensions and properties—otherwise the production cost would be prohibitive. The result is that the

Figure 5–2
Probability density curves for stress in an engineered system. (a) Variability of stresses in a relatively safe case. (b) Lower safety caused by overlap in stress distributions.

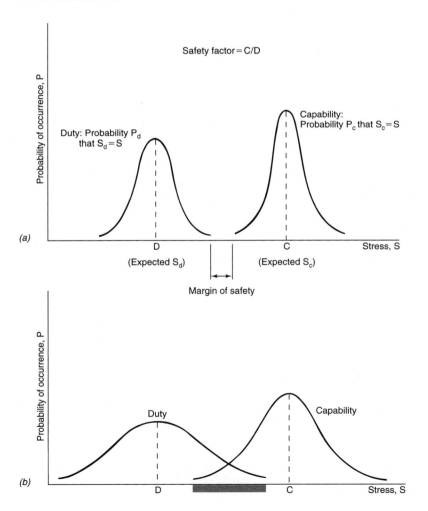

assembly's capability as a whole cannot be given by a single numerical value but must be expressed as a probability density that can be graphically depicted as a "capability" curve (Figure 5–2). For a given point on a capability curve, the value along the vertical axis gives the probability that the capability, or strength, is equal to the corresponding value along the horizontal axis.

A similar curve can be constructed for the duty that the assembly will actually experience. The stress exposure varies because of differences in loads, environmental conditions, or the manner in which the product is used. Associated with the capability and duty curves are nominal or, statistically speaking, expected values C and D. We often think and act only in terms of nominal or expected values. And with such a deterministic frame of mind, we may find it difficult to conceive of engineering as involving experimentation. The "safety factor" C/D rests comfortably in our consciences. But how sure can we be sure that our materials are truly close to their specified nominal properties, or that the loads will not vary too widely from their anticipated values or occur in environments hostile to the proper functioning of the materials?

At times the probability density curves of capability and, or, duty will take on flatter and broader shapes because of larger than expected variances as indicated in Figure 5–2b. If the respective values of D and C (shown on the horizontal axis for stress, S) remain the same, then so does the safety factor C/D. Now, however, there can be a pronounced overlap in the shaded region of the curves at worrisome values of probability. Edward B. Haugen has warned that "the safety factor concept completely ignores the facts of variability that result in different reliabilities for the same safety factor."[13]

A more appropriate measure of safety would be the "margin of safety," which is shown in Figure 5–2a. If it is difficult to compute such a margin of safety for ordinary loads used every day, imagine the added difficulties that arise when repeatedly changing loads have to be considered.

Risk-Benefit Analyses

Many large projects, especially public works, are justified on the basis of a risk-benefit analysis. The questions answered by such a study are the following: Is the product worth the risks connected with its use? What are the benefits? Do they outweigh the risks?

---

[13] Edward B. Haugen, *Probabilistic Approaches to Design* (New York: John Wiley & Sons, 1968), 5. Also see Haugen's *Probabilistic Mechanical Design* (New York: Wiley, 1980), 357.

We are willing to take on certain levels of risk as long as the project (activity, product, or system) promises sufficient benefit or gain. If risk and benefit can both be readily expressed in a common set of units (say, lives or dollars), it is relatively easy to carry out a risk-benefit analysis and to determine whether we can expect to come out on the benefit side. For example, an inoculation program may produce some deaths, but it is worth the risk if many more lives are saved by suppressing an imminent epidemic.

A closer examination of risk-benefit analyses reveals some conceptual difficulties.[14] Both risks and benefits lie in the future. As there is some uncertainty associated with them, we should address their expected values (provided such a model fits the situation); in other words, we should multiply the magnitude of the potential loss by the probability of its occurrence, and similarly with the gain. But who establishes these values, and how? If the benefits are about to be realized in the near future but the risks are far off, how is the future to be discounted in terms of, say, an interest rate so we can compare present values? What if the benefits accrue to one party, and the risks are incurred by another party?

The matter of delayed effects presents particular difficulties when an analysis is carried out during a period of high interest rates. Under such circumstances, the future is discounted too heavily because the very low present values of cost or benefit do not give a true picture of what a future generation will face.

How should one proceed when risks or benefits are composites of ingredients that cannot be added in a common set of units, as for instance in assessing effects on health plus aesthetics plus reliability? At most, one can compare designs that satisfy some constraints in the form of "dollars not to exceed X, health not to drop below Y" and try to compare aesthetic values with those constraints. Or when the risks can be expressed and measured in one set of units (say, deaths on the highway) and benefits in another (speed of travel), we can employ the ratio of risks to benefits for different designs when comparing the designs.

It should be noted that risk-benefit analysis, like cost-benefit analysis, is concerned with the advisability of undertaking a *project.* When we judge the relative merits of different *designs,* however, we move away from this concern. Instead, we are dealing with something similar to cost-effectiveness analysis, which asks what design has the greater merit, given that the project is actually to be carried out. Sometimes the shift from one type of

---

[14] See especially Matthew D. Adler and Eric A. Posner, eds., *Cost-Benefit Analysis* (Chicago: University of Chicago Press, 2001).

consideration to the other is so subtle that it passes unnoticed. Nevertheless, engineers should be aware of the differences so that they do not unknowingly carry the assumptions behind one kind of concern into their deliberations over the other.

These difficulties notwithstanding, there is a need in today's technological society for some commonly agreed-on process—or at least a process open to scrutiny and open to modification as needed—for judging the acceptability of potentially risky projects. What we must keep in mind is the following ethical question: "Under what conditions, if any, is someone in society entitled to impose a risk on someone else on behalf of a supposed benefit to yet others?"[15] Here we must not restrict our thoughts to average risks and benefits, but we should also consider those worst-case scenarios of persons exposed to maximum risks while they are also reaping only minimum benefits. Are their rights violated? Are they provided safer alternatives? In examining this problem further, we should also trace our steps back to an observation on risk perception made earlier: A risk to a known person (or to identifiable individuals) is perceived differently from statistical risks merely read or heard about. What this amounts to is that engineers do not affect just an amorphous public; their decisions have a direct impact on people who feel the impact acutely, and that fact should be taken into account just as seriously as are studies of statistical risk.

Personal Risk versus Public Risk

Given sufficient information, an individual can decide whether to participate in (or consent to exposure to) a risky activity (an experiment). Individuals are more ready to assume voluntary risks than they are involuntary risks, or activities over which they have no control, sometimes even when the voluntary risks are 1,000 times more likely to produce a fatality than the involuntary ones.[16]

The difficulty in assessing personal risks is magnified when we consider involuntary risks. Take John and Ann Smith and their discomfort over living near a refinery. Assume the general public was all in favor of building a new refinery at that location, and assume the Smiths already lived in the area. Would they and others in their situation have been justified in trying to veto its construction? Would they have been entitled to compensation if the plant was built over their objections anyway? If so, how much

---

[15] Council for Science and Society, *The Acceptability of Risks*, 37.

[16] Chauncey Starr, "Social Benefit versus Technological Risk," *Science* 165 (1969): 1232–38.

compensation would have been adequate? These questions arise in many cases, obviously including nuclear power plant siting.

The problem of quantification alone raises innumerable problems in assessing personal safety and risk. How, for instance, is one to assess the dollar value of an individual's life? This question is as difficult as deciding whose life is worth saving, should such a choice ever have to be made.

Some would advocate that the marketplace should decide, assuming market values can come into play. But today there is no over-the-counter trade in lives. Nor are even more mundane gains and losses easily priced. If the market is being manipulated, or if there is a wide difference between "product" cost and sales price, it matters under what conditions the buying or selling takes place. For example, if one buys a loaf of bread, it can matter whether it is just one additional daily loaf among others; it is different when it is the first loaf available in weeks. Or, if you are compensated for a risk by an amount based on the exposure tolerance of the average person, yet your tolerance of a condition or your propensity to be harmed is much greater than average, the compensation is apt to be inadequate.

The result of these difficulties in assessing personal risk is that analysts employ whatever quantitative measures are ready at hand. In regard to voluntary activities, one could possibly make judgments on the basis of the amount of life insurance a person buys. Is that individual going to offer the same amount to a kidnapper to be freed? Or is there likely to be a difference between future events (requiring insurance) and present events (demand for ransom)? In assessing a hazardous job, one might look at the increased wages a worker demands to carry out the task. Faced with the wide range of variables possible in such assessments, one can only suggest that an open procedure, overseen by trained arbiters, be employed in each case as it arises. Conversely, for people taken in a population-at-large context, it is much easier to use statistical averages without giving offense to anyone in particular.

Risks and benefits to the public at large are more easily determined because individual differences tend to even out as larger numbers of people are considered. Also, assessment studies relating to technological safety can be conducted more readily in the detached manner of a macroscopic view as statistical parameters take on greater significance. This occurs, for example, in studies by the National Highway Traffic Safety Administration (NHTSA), which proposes a value for human life based on loss of future income and other costs associated with an accident. Yet, studies repeatedly show that "policy analysts do not evaluate the risk of their subjects' lives as highly as people evaluate risks to their own (and others') lives. Consequently, too many risks are taken."[17]

Examples of Improved Safety

This is not a treatise on design; therefore, only a few simple examples will be given to show that safety need not rest on elaborate contingency features.

The first example is the magnetic door catch introduced on refrigerators to prevent death by asphyxiation of children accidentally trapped in them. The catch in use today permits the door to be opened from the inside without major effort. It also happens to be cheaper than the older types of latches.

The second example is the dead-man handle used by the engineer (engine driver) to control a train's speed. The train is powered only as long as some pressure is exerted on the handle. If the engineer becomes incapacitated and lets go of the handle, the train stops automatically. Perhaps cruise controls for newer-model automobiles should come equipped with a similar feature.

The motor-reversing system shown in Figure 5–3 gives still another example of a situation in which the introduction of a safety feature involves merely the proper arrangement of functions

Figure 5–3

Reversing switch for a permanent magnet motor. (a) Arms 1 and 2 of the switch are both raised by a solenoid (not shown). If either one does not move—say, a contact sticks—while the other does, there is a short across the battery. The battery will discharge and be useless even after the trouble is detected. (b) By exchanging the positions of battery and motor, a stuck switch will cause no harm to the battery. (The motor can be shorted without harm.)

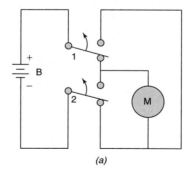 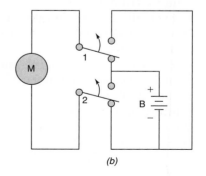

(a)   (b)

---

[17] Shulamit Kahn, "Economic Estimates of the Value of Life," *IEEE Technology and Society Magazine* (June 1986): 24–29. Reprinted in Albert Flores, ed., *Ethics and Risk Management in Engineering* (Boulder, CO: Westview Press, 1988).

at no additional expense. As the mechanism is designed in Figure 5–3a, sticky contacts could cause battery B to be shorted, thus making it unavailable for further use even after the contacts are coaxed loose. A simple reconnection of wires as shown in Figure 5–3b removes that problem altogether.

In the rush to bring a product into the market, safety considerations are often slighted. This would not be so much the case if the venture were regarded as an experiment—an experiment that is about to enter its active phase as the product comes into the hands of the user. Space flights were carried out with such an attitude, but more mundane ventures involve less obvious dangers, and therefore less attention is usually paid to safety. If moral concerns alone do not sway engineers and their employers to be more heedful of potential risks, then recent trends in product liability law should certainly do so.

Three Mile Island

As our engineered systems grow more complex, it becomes more difficult to operate them. As Charles Perrow argues, our traditional systems tended to incorporate sufficient slack, which allowed system aberrations to be corrected in a timely manner.[18] Nowadays, he points out, subsystems are so tightly coupled within more complex total systems that it is not possible to alter a course safely unless it can be done quickly and correctly.

Operator errors were the main cause of the nuclear reactor accident at Three Mile Island (TMI). In addition, there were inadequate provisions for evacuation of nearby populations. This lack of safe exit is found in too many of our amazingly complex systems.

Briefly, this is what happened.[19] At 4:00 a.m. on March 28, 1979, Unit TMI-2 was operating under full automatic control at 97 percent of its rated power output. For 11 hours a maintenance crew had been working on a recurring minor problem. Resin beads were used in several demineralizers (labeled 14 in Figure 5–4) to clean or "polish" the water on its way from the steam condenser (12) back to the steam generator (3).[20] Some beads

---

[18] Charles Perrow, *Normal Accidents: Living with High-Risk Technologies* (Princeton, NJ: Princeton University Press, 1999).

[19] Kemeny Commission Report, *Report of the President's Commission on the Accident at Three Mile Island* (New York: Pergamon Press, 1979); Daniel F. Ford, *Three Mile Island* (New York: Viking, 1982); John F. Mason, "The Technical Blow-by-Blow: An Account of the Three Mile Island Accident," *IEEE Spectrum* 16 (November 1979): 33–42.

[20] Mitchell Rogovin and George T. Frampton Jr., *Three Mile Island: A Report to the Commissioners and the Public*, vol. 1, Nuclear Regulatory Commission Special Inquiry Group, NUREG/CR-1250, Washington, DC (January 1980), 3. Diagram in text used with permission of Mitchell Rogovin.

clogged the resin pipe from a demineralizer to a tank in which the resin was regenerated. In flushing the pipe with water, perhaps a cupful of water backed up into an air line that provided air for fluffing the resin in its regeneration tank. But that air line was connected to the air system that also served the control mechanisms of the large valves at the outlet of the demineralizers. Thus it happened that these valves closed unexpectedly.

With water flow interrupted in the secondary loop (26), all but one of the condensate booster pumps turned off. That caused the main feedwater pumps (23) and the turbine (10) to shut down as well. In turn, an automatic emergency system started up the auxiliary feedwater pumps (25). But with the turbines inoperative, there was little outlet for the heat generated by the fission process in the reactor core. The pressure in the reactor rose to more than 2,200 pounds per square inch, opening a pressure-relief valve (7) and signaling a SCRAM, in which control rods are lowered into the reactor core to stop the main fission process.

The open valve succeeded in lowering the pressure, and the valve was readied to be closed. Its solenoid was de-energized, and the operators were so informed by their control-panel lights. But something went wrong: The valve remained open, contrary to what the control panel indicated. Apart from this failure everything else had proceeded automatically as it was supposed to. Everything, that is, except for one other serious omission: The auxiliary pumps (25) that had been started automatically could not supply the auxiliary feedwater because block valves (24) had inadvertently been left closed after maintenance work done on them two days earlier. Without feedwater in the loop (26), the steam generator (3) boiled dry. Now there was practically no heat removal from the reactor, except through the relief valve. Water was pouring out through it at the rate of 220 gallons per minute. The reactor had not yet cooled down, and even with the control rods shutting off the main fission reaction there would still be considerable heat produced by the continuing radioactive decay of waste products.

Loss of water in the reactor caused one of a group of pumps, positioned at 15, to start automatically; another one of these pumps was started by the operators to rapidly replenish the water supply for the reactor core. Soon thereafter, the full emergency core-cooling system went into operation in response to low reactor pressure. Low reactor pressure can promote the formation of steam bubbles that reduce the effectiveness of heat transfer from the nuclear fuel to the water. There is a pressurizer that is designed to keep the reactor water under pressure. (The relief valve sits atop this pressurizer.) The fluid level in the pressurizer was also used as an indirect—and the only—means of measuring the water level in the reactor.

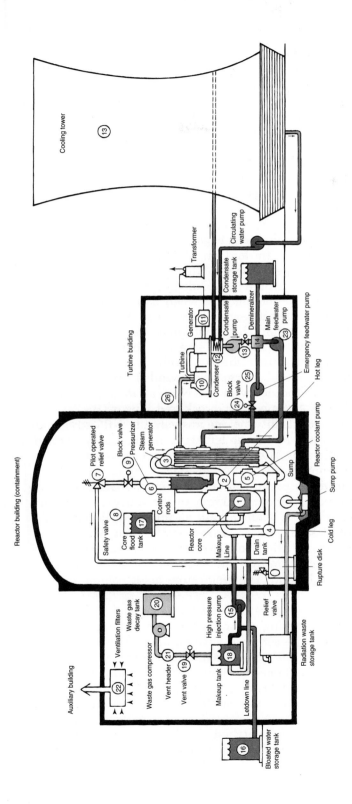

Figure 5–4

Schematic diagram of Three Mile Island nuclear power plant Unit 2. Pressurized water reactor system: Heat from reactor core (1) is carried away by water in a primary loop (1, 2, 3, 5, 4). In the steam generator (3) the heat is transferred to water in a secondary loop (26) at lower pres-sure. The secondary-loop water turns to steam in the steam generator or boiler (3), drives the turbine (10), turns into water in the condenser (12), and is circulated back to (3) by means of pumps (13, 23, and 25). (Adapted from John F. Mason, "The Technical Blow-by-Blow: An Account of the Three Mile Island Accident," *IEEE Spectrum*, 16 [November 1979], copyright © 1979 by the Institute of Electrical and Electronics Engineers, Inc., and from Mitchell Rogovin and George T. Frampton Jr., *Three Mile Island: A Report to the Commissioners and the Public*, vol. 1, Nuclear Regulatory Commission Special Inquiry Group, NUREG/CR=1250, Washington, DC [January 1980]).

The steam in the reactor vessel caused the fluid level in the pressurizer to rise. The operators, thinking they had resolved the problem and that they now had too much water in the reactor, shut down the emergency core-cooling system and all but one of the emergency pumps. Then they proceeded to drain water at a rate of 160 gallons per minute from the reactor, causing the pressure to drop. At this point they were still unaware of the water escaping through the open relief valve. Actually, they assumed some leakage, which occurred because of poor valve seating even under normal circumstances. It was this that made them disregard the high-temperature readings in the pipes (beyond location 7).

The steam bubbles in the reactor water covered much of the fuel, and the tops of the fuel rods began to crumble. The chemical reaction between the steam and Zircaloy covering the fuel elements produced hydrogen, some of which was released into the containment structure, where it exploded.

The situation was becoming dire when, two hours after the initial event, the next shift arrived for duty. With some fresh insights into the situation, the relief valve was deduced to be open. Blocking valve 9 in the relief line was then closed by the crew. Soon thereafter, with radiation levels in the containment building rising, a general alarm was sounded. Although there had been telephone contact with the Nuclear Regulatory Commission (NRC) as well as with Babcock and Wilcox (B&W), who had built the reactor facility, no one answered at NRC's regional office, and a message had to be left with an answering service. The fire chief of nearby Middletown was to hear about the emergency on the evening news.

In the meantime, a pump was transferring the drained water from the main containment building to the adjacent auxiliary building but not into a holding tank as intended; because of a blown rupture disk, the water landed on the floor. When there was indication of sufficient airborne radiation in the control room to force evacuation, all but essential personnel wearing respirators stayed behind. The respirators made communication difficult.

Eventually the operators decided to turn the high-pressure injection pumps on again, as the automatic system had been set to do all along. The core was covered once more with water, although there were still some steam and hydrogen bubbles on the loose. Thirteen and one-half hours after the start of the episode, there was finally hope of getting the reactor under control. Confusion over the actual state of affairs, however, continued for several days.

Nationwide, the public watched television coverage in disbelief as responsible agencies displayed their lack of emergency preparedness at both the reactor site and evacuation-planning centers. Years later one still reads about the steadily accumulating costs of decommissioning (defueling, decontaminating, entombing) Unit 2 at TMI, $1 billion so far, of which one-third is passed on to ratepayers—all this for what cost $700 million to build. Three Mile Island was a financial disaster and a blow to the reputation of the industry, but fortunately radioactive release was low, and cancer rates downwind are reported to be only slightly higher than normal. The reactor cleanup started in August 1979 and ended in December 1993 at a cost of $975 million. The other reactor (TMI-1) was restarted in 1985 and subsequently changed ownership.

Safe Exits

It is almost impossible to build a completely safe product or one that will never fail. The best one can do is to assure that when a product fails, (1) it will fail safely, (2) the product can be abandoned safely, or—at least—(3) the user can safely escape the product. Let us refer to these three conditions as *safe exit*. It is not obvious who should take the responsibility for providing safe exit. But apart from questions of who will build, install, maintain, and pay for a safe exit system there remains the crucial question of who will recognize the need for a safe exit.

Providing for a safe exit is an integral part of the social-experimental procedure—in other words, of sound engineering. The experiment is to be carried out without causing bodily or financial harm. If safety is threatened, the experiment must be terminated. The full responsibility cannot fall on the shoulders of a lone engineer, but one can expect the engineer to issue warnings when a safe exit does not exist or the experiment must be terminated. The only way one can justify the continuation of an experiment without safe exit is for all participants (including the subjects of the experiment) to have given valid consent for its continuation.

Here are some examples of what this might involve. Ships need lifeboats with enough spaces for all passengers and crew members. Buildings need usable fire escapes. The operation of nuclear power plants calls for realistic ways to evacuate nearby communities. The foregoing are examples of safe exits for people. Provisions are also needed for safe disposal of dangerous products and materials: Altogether too many truck accidents and train derailments have exposed communities to toxic gases, and too many dumps have let toxic wastes get to the groundwater table or into the hands of children. Finally, avoiding system failure might require redundant or alternative means of continuing a process when the original process fails. Examples would be backup systems for computer-based data banks, air traffic control systems, automated medical treatment systems, or sources of water for fire fighting.

Apart from a safety conscious design and thorough testing of any potentially dangerous product before it is delivered for use, it is of course necessary that its user have in place procedures for regular maintenance and safety checks. Beyond such measures there should also be in place (a) avenues for employees to freely report hazardous conditions regarding the design or the operation of the product without having to resort to whistle-blowing, and (b) emergency procedures based on human factors engineering that takes into account how people react and interact under conditions of stress as occurred during the TMI accident.

Discussion Questions

1. A worker accepts a dangerous job after being offered an annual bonus of $2,000. The probability that the worker may be killed in any one year is 1 in 10,000. This is known to the worker. The bonus may therefore be interpreted as a self-assessment of life with a value equal to $2,000 divided by 1/10,000, or $20 million. Is the worker more or less likely to accept the job if presented with the statistically nearly identical figures of a $100,000 bonus over 50 years (neglecting interest) and a 1/200 probability of a fatal accident during that period?

2. "Airless" paint spray guns do not need an external source of compressed air connected to the gun by a heavy hose (although they do need a cord to attach them to a power source) because they have incorporated a small electric motor and pump. One common design uses an induction motor that does not cause sparking because it does not require a commutator and brushes (which are sources of sparking). Nevertheless the gun carries a label warning users that electrical devices operated in paint spray environments pose special dangers. Another type of gun that, like the first, also requires only a power cord is designed to weigh less by using a high-speed universal motor and a disk-type pump. The universal motor does require a commutator and brushes, which cause sparking. This second kind of spray gun carries a warning similar to that attached to the first, but it states in addition that the gun should never be used with paints that employ highly volatile and flammable thinners such as naphtha. The instruction booklet is quite detailed in its warnings.

   A painter had been lent one of the latter types of spray guns. To clean the apparatus, he partially filled it with paint thinner and operated it. It caught fire, and the painter was severely burned as the fire spread. The instruction booklet was in the cardboard box in which the gun was kept, but it had not been read by the painter, who was a recent immigrant and did not read English very well. He had, however, used the first type of airless paint spray gun in a similar manner without mishap. The warning messages on both guns looked pretty much the same. Do you see any ethical problems in continuing over-the-counter sales of this second type of spray gun? What should the manufacturer of this novel, lightweight device do?

In answering these questions, consider the fact that courts have ruled that hidden design defects are not excused by warnings attached to the defective products or posted in salesrooms. Informed consent must rest on a more thorough understanding than can be transmitted to buyers by warning labels.

3. It has been said that Three Mile Island showed us the risks of nuclear power and the Arab oil embargo the risk of having no energy. Forcing hazardous products or services from the market has been criticized as closing out the options of those individuals or countries with rising aspirations who can now afford them and who may all along have borne more than their share of the risks without any of the benefits. Finally, pioneers have always exposed themselves to risk. Without risk there would be no progress. Discuss this problem of "the risk of no risk."[21]

4. Discuss the notion of safe exit, using evacuation plans for communities near nuclear power plants or chemical process plants.

5. Research the events at Chernobyl in 1986, and discuss what you see as the main similarities and differences with Three Mile Island.

---

[21] Aaron Wildavsky, "No Risk Is the Highest Risk of All," in *Ethical Problems in Engineering*, 2nd ed., ed. Albert Flores (Troy, NY: Rensselaer Polytechnic Institute 1980), 221–26.

CHAPTER 6

Workplace Responsibilities and Rights

Data General Corporation grew spectacularly during its first decade of operation, quickly becoming a Fortune 500 company that was ranked third in overall sales of small computers. However, it began to fall behind the competition and desperately needed a powerful new microcomputer to sustain its share of the market. The development of that computer is chronicled by Tracy Kidder in his Pulitzer Prize–winning book *The Soul of a New Machine*.

Tom West, one of Data General's most trusted engineers, convinced management that he could build the new computer within one year—an unprecedented time for a project of its importance. West assembled a team of fifteen exceptionally motivated although relatively inexperienced young engineers, many of whom were just out of school. Within six months they designed the central processing unit, and they delivered the complete computer ahead of schedule. Named the Eclipse MV/8000, the computer immediately became a major marketing success.

The remarkable success was possible in part because the engineers came to identify themselves with the project and the product: "Ninety-eight percent of the thrill comes from knowing that the thing you designed works, and works almost the way you expected it would. If that happens, part of you is in that machine."[1] The "soul" of the new machine was not any one person. Instead, it was the team of engineers who invested themselves in the product through their personal commitment to work together creatively with colleagues as part of a design group. As might be expected, personality clashes occurred during the sometimes frenzied work schedule, but conflict was minimized by a commitment to teamwork, collegiality, and shared identification with

---

[1] Tracy Kidder, *The Soul of a New Machine* (New York: Avon Books, 1981), 273.

the group's project. More worrisome, there were times when the engineers pushed themselves to their limits, imposing burdens on their families and their health, but for the most part those times remained limited.

Kidder ends his book by quoting a regional sales manager speaking to the sales representatives preparing to market the new computer: "'What motivates people?' he asked. He answered his own question, saying, 'Ego and the money to buy things that they and their families want.'"[2] The engineers, of course, cared about money and ego, but Kidder makes it clear that those motives could not explain how it was possible for them to accomplish what they did. Professionalism involves much more, including both a sense of fun and excitement, personal commitments that have moral dimensions, and teamwork.

The kind of commitments shown by the engineers understandably ranks high on the list of expectations that employers have of the engineers they employ or engage as consultants. Engineers in turn should see top performance at a professional level as their main responsibility, accompanied by others such as maintaining confidentiality and avoiding conflicts of interest. Engineers also need the opportunity to perform responsibly, and this means that their professional and employee rights must be respected.

6.1 Confidentiality and Conflicts of Interest

Confidentiality: Definition

The duty of confidentiality is the duty to keep secret all information deemed desirable to keep secret. Deemed by whom? Basically, it is any information that the employer or client would like to have kept secret to compete effectively against business rivals. Often this is understood to be any data concerning the company's business or technical processes that are not already public knowledge. Although this criterion is somewhat vague, it clearly points to the employer or client as the main source of the decision as to what information is to be treated as confidential.

"Keep secret" is a relational expression. It always makes sense to ask, "Secret with respect to whom?" In the case of some government organizations, such as the Federal Bureau of Investigation (FBI) and Central Intelligence Agency (CIA), highly elaborate systems for classifying information have been developed that identify which individuals and groups may have access to what information. Within other governmental agencies and private companies, engineers and other employees are usually expected to withhold information labeled "confidential" from unauthorized people both inside and outside the organization.

---

[2] Ibid., 291, italics omitted.

Several related terms should be distinguished. *Privileged information* literally means "available only on the basis of special privilege," such as the privilege accorded an employee working on a special assignment. *Proprietary information* is information that a company owns or is the proprietor of, and hence is a term carefully defined by property law. A rough synonym for "proprietary information" is *trade secret,* which can be virtually any type of information that has not become public, which an employer has taken steps to keep secret, and which is thereby given limited legal protection in common law (law generated by previous court rulings) that forbids employees from divulging it. *Patents* legally protect specific products from being manufactured and sold by competitors without the express permission of the patent holder. Trade secrets have no such protection, and a corporation can learn about a competitor's trade secrets through legal means— for instance, "reverse engineering," in which an unknown design or process can be traced out by analyzing the final product. But patents do have the drawback of being public and thus allowing competitors an easy means of working around them by finding alternative designs.

Confidentiality and Changing Jobs

The obligation to protect confidential information does not cease when employees change jobs. If it did, it would be impossible to protect such information. Former employees would quickly divulge it to their new employers or, perhaps for a price, sell it to competitors of their former employers. Thus, the relationship of trust between employer and employee in regard to confidentiality continues beyond the formal period of employment. Unless the employer gives consent, former employees are barred indefinitely from revealing trade secrets. This provides a clear illustration of the way in which the professional integrity of engineers involves much more than mere loyalty to one's present employer.

Yet thorny problems arise in this area. Many engineers value professional advancement more than long-term ties with any one company, and so they change jobs frequently. Engineers in research and development are especially likely to have high rates of turnover. They are also the people most likely to be exposed to important new trade secrets. Moreover, when they transfer into new companies they often do the same kind of work as before— precisely the type of situation in which trade secrets of their old companies may have relevance, a fact that could have strongly contributed to their having readily found new employment.

A high-profile case of trade secret violations was settled in January 1997 (without coming to trial) when Volkswagen AG (VW) agreed to pay General Motors Corporation (GM) and its

German subsidiary Adam Opel $100 million in cash and to buy
$1 billion in parts from GM over the next seven years. Why?
Because in March 1993, Jose Ignacio Lopez, GM's highly effective
manufacturing expert, left GM to join VW, a fierce competitor in
Europe, and took with him not only three colleagues and know-
how, but also copies of confidential GM documents.

A more legally important case concerned Donald Wohlgemuth,
a chemical engineer who at one time was manager of B.F.
Goodrich's space suit division.[3] Technology for space suits was
undergoing rapid development, with several companies competing for government contracts. Dissatisfied with his salary and
the research facilities at B.F. Goodrich, Wohlgemuth negotiated
a new job with International Latex Corporation as manager of
engineering for industrial products. International Latex had
just received a large government subcontract for developing the
Apollo astronauts' space suits, and that was one of the programs
Wohlgemuth would manage.

The confidentiality obligation forbid Wohlgemuth from revealing any trade secrets of Goodrich to his new employer. This was
easier said than done. Of course it is possible for employees in his
situation to refrain from explicitly stating processes, formulas,
and material specifications. Yet in exercising their general skills
and knowledge, it is virtually inevitable that some unintended
"leaks" will occur. An engineer's knowledge base generates an
intuitive sense of what designs will or will not work, and trade
secrets form part of this knowledge base. To fully protect the
secrets of an old employer on a new job would thus virtually
require that part of the engineer's brain be removed.

Is it perhaps unethical, then, for employees to change jobs in
cases where unintentional revelations of confidential information are a possibility? Some companies have contended that it is.
Goodrich, for example, charged Wohlgemuth with being unethical
in taking the job with International Latex. Goodrich also went to
court seeking a restraining order to prevent him from working for
International Latex or any other company that developed space
suits. The Ohio Court of Appeals refused to issue such an order,
although it did issue an injunction prohibiting Wohlgemuth from
revealing any Goodrich trade secrets. Their reasoning was that
although Goodrich had a right to have trade secrets kept confidential, it had to be balanced against Wohlgemuth's personal
right to seek career advancement. And this would seem to be the
correct moral verdict as well.

---

[3] Michael S. Baram, "Trade Secrets: What Price Loyalty?" *Harvard Business
Review* (November–December 1968). Reprinted in *Ethical Issues in Engineering*,
ed. Deborah G. Johnson (Englewood Cliffs, NJ: Prentice Hall, 1991), 279–90.

Confidentiality and Management Policies

What might be done to recognize the legitimate personal interests and rights of engineers and other employees while also recognizing the rights of employers in this area?[4] One approach is to use employment contracts that place special restrictions on future employment. Traditionally, those restrictions centered on the geographical location of future employers, the length of time after leaving the present employer before one can engage in certain kinds of work, and the type of work it is permissible to do for future employers. Thus, Goodrich might have required as a condition of employment that Wohlgemuth sign an agreement that if he sought work elsewhere he would not work on space suit projects for a competitor in the United States for five years after leaving Goodrich.

Yet such contracts are hardly agreements between equals, and they threaten the right of individuals to pursue their careers freely. For this reason the courts have tended not to recognize such contracts as binding, although they do uphold contractual agreements forbidding the disclosure of trade secrets.

A different type of employment contract is perhaps not so threatening to employee rights in that it offers positive benefits in exchange for the restrictions it places on future employment. Consider a company that normally does not have a portable pension plan. It might offer such a plan to an engineer in exchange for an agreement not to work for a competitor on certain kinds of projects for a certain number of years after leaving the company. Or another clause might offer an employee a special postemployment annual consulting fee for several years on the condition that he or she not work for a direct competitor during that period.

Other tactics aside from employment contract provisions have been attempted by various companies. One is to place tighter controls on the internal flow of information by restricting access to trade secrets except where absolutely essential. The drawback to this approach is that it may create an atmosphere of distrust in the workplace. It might also stifle creativity by lessening the knowledge base of engineers involved in research and development.

One potential solution is for employers to help generate a sense of professional responsibility among their staff that reaches beyond merely obeying the directives of current employers. Engineers can then develop a real sensitivity to the moral conflicts they may be exposed to by making certain job changes. They can arrive at a greater appreciation of why trade secrets are important in a competitive system, and they can learn to take

---

[4] Ibid., 285–90.

the steps necessary to protect them. In this way, professional concerns and employee loyalty can become intertwined and reinforce each other.

Confidentiality: Justification

On what moral basis does the confidentiality obligation rest, with its wide scope and obvious importance? The primary justification is to respect the autonomy (freedom, self-determination) of individuals and corporations and to recognize their legitimate control over some private information concerning themselves.[5] Without that control, they could not maintain their privacy and protect their self-interest insofar as it involves privacy. Just as patients should be allowed to maintain substantial control over personal information, so employers should have some control over the private information about their companies. All the major ethical theories recognize the importance of autonomy, whether it is understood in terms of rights to autonomy, duties to respect autonomy, the utility (as in utilitarian ethics) of protecting autonomy, or the virtue of respect for others.

Additional justifications include trustworthiness: Once practices of maintaining confidentiality are established socially, trust and trustworthiness can grow. Thus, when clients go to attorneys or tax accountants they expect them to maintain confidentiality, and the professional indicates that confidentiality will be maintained. Similarly, employees often make promises (in the form of signing contracts) not to divulge certain information considered sensitive by the employer.

In addition, there are public benefits in recognizing confidentiality relationships within professional contexts. For example, if patients are to have the best chances of being cured, they must feel completely free to reveal the most personal information about themselves to physicians, and that requires trust that the physician will not divulge private information. Likewise, the economic benefits of competitiveness within a free market are promoted when companies can maintain some degree of confidentiality about their products. Developing new products often requires investing large resources in acquiring new knowledge. The motivation to make those investments might diminish if that knowledge were immediately dispersed to competitors who could then quickly make better products at lesser cost, as they did not have to make comparable investments in research and development.

Confidentiality has its limits, particularly when it is invoked to hide misdeeds. Investigations into a wide variety of white-collar

---

[5] Sissela Bok, *Secrets* (New York: Pantheon Books, 1982), 116–35.

crimes covered up by management in industry or public agencies have been thwarted by invoking confidentiality or false claims of secrecy based on national interest. And the possibility of justified whistle-blowing, discussed in Chapter 7, raises additional complications.

Conflicts of Interest: Definition and Examples

We turn now to some equally thorny issues concerning conflicts of interest. Professional conflicts of interest are situations where professionals have an interest that, if pursued, might keep them from meeting their obligations to their employers or clients. Sometimes such an interest involves serving in some other professional role, say, as a consultant for a competitor's company. Other times it is a more personal interest, such as making substantial private investments in a competitor's company.

Concern about conflicts of interest largely centers on their potential to distort good judgment in faithfully serving an employer or client.[6] Exercising good judgment means arriving at beliefs on the basis of expertise and experience, as opposed to merely following simple rules. Thus, we can refine our definition of conflicts of interest by saying that they typically arise when *two* conditions are met: (1) The professional is in a relationship or role that requires exercising good judgment on behalf of the interests of an employer or client, and (2) the professional has some additional or side interest that could threaten good judgment in serving the interests of the employer or client—either the good judgment of that professional or the judgment of a typical professional in that situation. Why the reference to "a typical professional"? There might be conclusive evidence that the actual persons involved would never allow a side interest to affect their judgment, yet they are still in a conflict of interest.

"Conflict of interest" and "conflicting interests" are not synonyms.[7] A student, for example, may have interests in excelling on four final exams. She knows, however, that there is time to study adequately for only three of them, and so she must choose which interest not to pursue. In this case "conflicting interests" means a person has two or more desires that cannot all be satisfied given the circumstances. But there is no suggestion that it is morally wrong or problematic to try pursuing them all. By

---

[6] Michael Davis, "Conflict of Interest," *Business and Professional Ethics Journal* 1 (Summer 1982): 17–27; Paula Wells, Hardy Jones, and Michael Davis, *Conflicts of Interest in Engineering* (Dubuque, IA: Kendall/Hunt, 1986); Michael Davis and Andrew Stark, eds., *Conflict of Interest in the Professions* (New York: Oxford University Press, 2001).

[7] Joseph Margolis, "Conflict of Interest and Conflicting Interests," in *Ethical Theory and Business,* ed. T. Beauchamp and N. Bowie (Englewood Cliffs, NJ: Prentice Hall, 1979), 361.

contrast, in professional conflicts of interest it is often physically or economically possible to pursue all of the conflicting interests but doing so would be morally problematic.

Because of the great variety of possible outside interests, conflicts of interest can arise in innumerable ways, and with many degrees of subtlety. We will sample only a few of the more common situations involving (1) gifts, bribes, and kickbacks, (2) interests in other companies, and (3) insider information.

Gifts, Bribes, and Kickbacks. A bribe is a substantial amount of money or goods offered beyond a stated business contract with the aim of winning an advantage in gaining or keeping the contract, and where the advantage is illegal or otherwise unethical.[8] *Substantial* is a vague term, but it alludes to amounts, beyond acceptable gratuities, that are sufficient to distort the judgment of a typical person. Typically, although not always, bribes are made in secret. Gifts are not bribes as long as they are small gratuities offered in the normal conduct of business. Prearranged payments made by contractors to companies or their representatives in exchange for contracts actually granted are called kickbacks. When suggested by the granting party to the party bidding on the contract, the latter often defends its participation in such an arrangement as having been subjected to "extortion."

Often, companies give gifts to selected employees of government agencies or partners in trade. Many such gifts are unobjectionable, some are intended as bribes, and still others create conflicts of interest that do not, strictly speaking, involve bribes. What are the differences? In theory, these distinctions may seem clear, but in practice they become blurry. Bribes are illegal or immoral because they are substantial enough to threaten fairness in competitive situations, whereas gratuities are of smaller amounts. Some gratuities play a legitimate role in the normal conduct of business, whereas others can bias judgment like a bribe does. Much depends on the context, and there are many gray areas, which is why companies often develop elaborate guidelines for their employees.

What about gifts in routine business contexts? Is it all right to accept the occasional luncheon paid for by vendors giving sales presentations, or a gift one believes is given in friendship rather than for influence? Codes of ethics sometimes take a hard line in forbidding all such gratuities, but many employers set forth more flexible policies. Company policies generally ban any gra-

---

[8] Cf. Michael S. Pritchard, "Bribery: The Concept," *Science and Engineering Ethics* 4, no. 3 (1998): 281–86.

tuities that have more than nominal value or exceed widely and openly accepted normal, business practice. An additional rule of thumb is: "If the offer or acceptance of a particular gift could have embarrassing consequences for your company if made public, then do not accept the gift."

Interests in Other Companies. Some conflicts of interest consist in having an interest in a competitor's or a subcontractor's business. One blatant example is actually working for the competitor or subcontractor as an employee or consultant. Another example is partial ownership or substantial stockholdings in the competitor's business. Does holding a few shares of stock in a company one has occasional dealings with constitute a conflict of interest? Usually not, but as the number of shares of stock increases, the issue becomes blurry. Again, is there a conflict of interest if one's spouse works for a subcontractor to one's company? Usually not, but a conflict of interest arises if one's job involves granting contracts to that subcontractor.

Should there be a general prohibition on *moonlighting,* that is, working in one's spare time for another company? That would violate the rights to pursue one's legitimate self-interest. Moonlighting usually creates conflicts of interest only in special circumstances, such as working for competitors, suppliers, or customers. Even then, in rare situations, an employer sometimes gives permission for exceptions, as for example when the experience gained would greatly promote business interests. A special kind of conflict of interest arises, however, when moonlighting leaves one exhausted and thereby harms job performance.[9]

Conflicts of interest arise in academic settings as well. For example, a professor of electrical engineering at a West Coast university was found to have used $144,000 in grant funds to purchase electronic equipment from a company he owned in part. He had not revealed his ownership to the university, he had priced the equipment much higher than market value, and some of the purchased items were never received. The Supplier Information Form and Sole Source Justification Statements had been submitted as required but with falsified content. In addition, the professor had hired a brother and two sisters for several years, concealing their relationship to him in violation of antinepotism rules and paying them for research work they did not perform. All told, he had defrauded the university of at least $500,000 in research funds. Needless to say, the professor lost his university

---

[9] George L. Reed, "Moonlighting and Professional Responsibility," *Journal of Professional Activities: Proceedings of the American Society of Civil Engineers* 96 (September 1970): 19–23.

position and had to stand trial in civil court when an internal audit and subsequent hearings revealed these irregularities.

Insider Information. An especially sensitive conflict of interest consists in using "inside" information to gain an advantage or set up a business opportunity for oneself, one's family, or one's friends. The information might concern one's own company or another company with which one does business. For example, engineers might tell their friends about the impending announcement of a revolutionary invention, which they have been perfecting, or of their corporation's plans for a merger that will greatly improve the worth of another company's stock. In doing so, they give those friends an edge on an investment promising high returns. Owning stock in the company for which one works is of course not objectionable, and this is often encouraged by employers. But that ownership should be based on the same information available to the general public.

Moral Status of Conflicts of Interest

What is wrong with employees having conflicts of interest? Most of the answer is obvious from our definition: Employee conflicts of interest occur when employees have interests that if pursued could keep them from meeting their obligations to serve the interests of the employer or client for whom they work. Such conflicts of interest should be avoided because they threaten to prevent one from fully meeting those obligations.

More needs to be said, however. Why should mere threats of possible harm always be condemned? Suppose that substantial good might sometimes result from pursuing a conflict of interest?

In fact, it is not always unethical to pursue conflicts of interest. In practice, some conflicts are thought to be unavoidable, or even acceptable. One illustration of this is that the government allows employees of aircraft manufacturers, such as Boeing or McDonnell Douglas, to serve as government inspectors for the Federal Aviation Agency (FAA). The FAA is charged with regulating airplane manufacturers and making objective safety and quality inspections of the airplanes they build. Naturally the dual roles—government inspector and employee of the manufacturer being inspected—could bias judgments. Yet with careful screening of inspectors, the likelihood of such bias is said to be outweighed by the practical necessities of airplane inspection. The options would be to greatly increase the number of nonindustry government workers (at great expense to taxpayers) or to do without government inspection altogether (putting public safety at risk).

Even when conflicts of interest are unavoidable or reasonable, employees are still obligated to inform their employers and obtain approval. This suggests a fuller answer to why conflicts of interest are generally prohibited: (1) The professional obligation to employers is very important in that it overrides in the vast majority of cases any appeal to self-interest on the job, and (2) the professional obligation to employers is easily threatened by self-interest (given human nature) in a way that warrants especially strong safeguards to ensure that it is fulfilled by employees.

Many conflicts of interest violate trust, in addition to undermining specific obligations. Employed professionals are in fiduciary (trust) relationships with their employers and clients. Allowing side interests to distort one's judgment violates that trust. And additional types of harm can arise as well. Many conflicts of interest are especially objectionable in business affairs precisely because they pose risks to free competition. In particular, bribes and large gifts are objectionable because they lead to awarding contracts for reasons other than the best work for the best price.

As a final point, we should note that even the appearance of conflicts of interest, especially appearances of seeking a personal profit at the expense of one's employer, is considered unethical because the appearance of wrongdoing can harm a corporation as much as any actual bias that might result from such practices.

Discussion Questions

1. Consider the following example:

> Who owns your knowledge? Ken is a process engineer for Stardust Chemical Corp., and he has signed a secrecy agreement with the firm that prohibits his divulging information that the company considers proprietary.
> Stardust has developed an adaptation of a standard piece of equipment that makes it highly efficient for cooling a viscous plastics slurry. (Stardust decides not to patent the idea but to keep it as a trade secret.) Eventually, Ken leaves Stardust and goes to work for a candy-processing company that is not in any way in competition. He soon realizes that a modification similar to Stardust's trade secret could be applied to a different machine used for cooling fudge and, at once, has the change made.[10]

Has Ken acted unethically?

---

[10] Philip M. Kohn and Roy V. Hughson, "Perplexing Problems in Engineering Ethics," *Chemical Engineering* 87 (May 5, 1980): 102. Quotations used with permission of McGraw-Hill Book Co.

2. In the following case, are the actions of Client A morally permissible?

> Client A solicits competitive quotations on the design and construction of a chemical plant facility. All the bidders are required to furnish as a part of their proposals the processing scheme planned to produce the specified final products. The process generally is one which has been in common use for several years. All of the quotations are generally similar in most respects from the standpoint of technology.
>
> Contractor X submits the highest-price quotation. He includes in his proposals, however, a unique approach to a portion of the processing scheme. Yields are indicated to be better than current practice, and quality improvement is apparent. A quick laboratory check indicates that the innovation is practicable.
>
> Client A then calls on Contractor Z, the low bidder, and asks him to evaluate and bid on an alternate scheme conceived by Contractor X. Contractor Z is not told the source of alternative design. Client A makes no representation in his quotation request that replies will be held in confidence.[11]

3. American Potash and Chemical Corporation advertised for a chemical engineer having industrial experience with titanium oxide. It succeeded in hiring an engineer who had formerly supervised E. I. DuPont de Nemours and Company's production of titanium oxide. DuPont went to court and succeeded in obtaining an injunction prohibiting the engineer from working on American Potash's titanium oxide projects. The reason given for the injunction was that it would be inevitable that the engineer would disclose some of DuPont's trade secrets.[12] Defend your view as to whether the court injunction was morally warranted or not.

4. Engineer Doe is employed on a full-time basis by a radio broadcast equipment manufacturer as a sales representative. In addition, Doe performs consulting engineering services to organizations in the radio broadcast field, including analysis of their technical problems and, when required, recommendation of certain radio broadcast equipment as may be needed. Doe's engineering reports to his clients are prepared in form for filing with the appropriate governmental body having jurisdiction over radio broadcast facilities. In some cases Doe's engineering reports recommend the use of broadcast equipment manufac-

---

[11] Philip L. Alger, N. A. Christensen, and Sterling P. Olmsted, *Ethical Problems in Engineering* (New York: John Wiley & Sons, 1965), 111. Quotations used with permission of the publisher.

[12] Charles M. Carter, "Trade Secrets and the Technical Man," *IEEE Spectrum* 6 (February 1969): 54.

tured by his employer. Can Doe ethically provide consulting services as described?[13]

5. Henry is in a position to influence the selection of suppliers for the large volume of equipment that his firm purchases each year. At Christmas time, he usually receives small tokens from several salesmen, ranging from inexpensive ballpoint pens to a bottle of liquor. This year, however, one salesman sends an expensive briefcase stamped with Henry's initials.[14]

Should Henry accept the gift? Should he take any further course of action?

6.2 Teamwork and Rights

Working effectively as an engineer requires the virtues of loyalty to employers and organizations, collegiality, respect for authority, and contributing to an ethical climate within the organization. It also involves respect for the rights of engineers and others who work together to achieve common goals.

An Ethical Corporate Climate

An ethical climate is a working environment that is conducive to morally responsible conduct. Within corporations it is produced by a combination of formal organization and policies, informal traditions and practices, and personal attitudes and commitments. Engineers can make a vital contribution to such a climate, especially as they move into technical management and then more general management positions.

Professionalism in engineering would be threatened at every turn in a corporation devoted primarily to powerful egos. Sociologist Robert Jackall describes several such corporations in his book *Moral Mazes* as organizations that reduce (and distort) corporate values to merely following orders: "What is right in the corporation is what the guy above you wants from you. That's what morality is in the corporation."[15] Jackall describes a world in which professional standards are disregarded by top-level managers preoccupied with maintaining self-promoting images and forming power alliances with other managers. Hard work, commitment to worthwhile and safe products, and even profit-making take a back seat to personal survival in the tumultuous world of corporate takeovers and layoffs. It is noteworthy that Jackall's book is based primarily on his study of several large

---

[13] *NSPE Opinions of the Board of Ethical Review,* Case 75.10, National Society of Professional Engineers, Washington, DC, www.nspe.org.

[14] Philip M. Kohn and Roy V. Hughson, "Perplexing Problems in Engineering Ethics," *Chemical Engineering* 87 (May 5, 1980): 104. Quotations in text used with permission of McGraw-Hill Book Co.

[15] Robert Jackall, *Moral Mazes: The World of Corporate Managers* (New York: Oxford University Press, 1988), 6, 109 (italics removed).

chemical and textile companies during the 1980s, companies notorious for indifference to worker safety (including cotton-dust poisoning) and environmental degradation (especially chemical pollution).

What are the defining features of an ethical corporate climate? There are at least four. First, ethical values in their full complexity are widely acknowledged and appreciated by managers and employees alike. Responsibilities to all constituencies of the corporation are affirmed—not only to stockholders, but also to customers, employees, and all other stakeholders in the corporation.

Second, the use of ethical language is honestly applied and recognized as a legitimate part of corporate dialogue. One way to emphasize this legitimacy is to make prominent a corporate code of ethics. Another way is to explicitly include a statement of ethical responsibilities in the job descriptions of all layers of management.

Third, top management sets a moral tone in words, in policies, and by personal example. Official pronouncements asserting the importance of professional conduct in all areas of the corporation must be backed by support for professionals who work according to the guidelines outlined in professional codes of ethics. Whether or not there are periodic workshops on ethics or formal brochures on social responsibility distributed to all employees, what is most important is fostering confidence that management is serious about ethics.

Fourth, there are procedures for conflict resolution. One avenue is to create ombudspersons or designated executives with whom employees can have confidential discussions about moral concerns. Equally important is educating managers about conflict resolution. There are also ties of loyalty and collegiality that help minimize conflicts in the first place.

Loyalty and Collegiality

Loyalty to an employer can mean two things.[16] *Agency-loyalty* is acting to fulfill one's contractual duties to an employer. These duties are specified in terms of the particular tasks for which one is paid, as well as the more general activities of cooperating with colleagues and following legitimate authority within the corporation. As its name implies, agency-loyalty is entirely a matter

---

[16] John Ladd, "Loyalty," in *The Encyclopedia of Philosophy,* vol. 4, ed. Paul Edwards (New York: Macmillan, 1967), 97–98. Also see Marcia Baron, *The Moral Status of Loyalty* (Dubuque, IA: Kendall/Hunt, 1984); and John H. Fielder, "Organizational Loyalty," *Business and Professional Ethics Journal* 11 (Spring 1992): 83.

of actions, such as doing one's job and not stealing from one's employer, regardless of the motives for it.

Attitude-loyalty, by contrast, has as much to do with attitudes, emotions, and a sense of personal identity as it does with actions. It can be understood as agency-loyalty that is motivated by a positive identification with the group to which one is loyal. It implies seeking to meet one's moral duties to a group or organization willingly, with personal attachment and affirmation, and with a reasonable degree of trust. People who do their work grudgingly or spitefully are not loyal in this sense, even though they may adequately perform all their work responsibilities and hence manifest agency-loyalty.

When codes of ethics assert that engineers ought to be loyal (or faithful) to employers, is agency-loyalty or attitude-loyalty meant? Within proper limits, agency-loyalty to employers is an obligation, or rather it comprises the sum total of obligations to employers to serve the corporation in return for the contractual benefits from the corporation. But it is not the sole or paramount obligation of engineers. According to the National Society of Professional Engineers (NSPE) Code of Ethics, and many other codes, the overriding obligation of engineers remains to "hold paramount the safety, health and welfare of the public."

What about attitude-loyalty: Is it obligatory? In our view, attitude-loyalty is often a virtue but not strictly an obligation. It is good when it contributes to a sense of corporate community and, thereby, increases the prospects for corporations to meet their desirable goals of productivity. We might say that loyalty is a "dependent virtue": its desirability depends on its contribution to *valuable* projects and communities to which it contributes.[17]

Any discussion of employee loyalty must address the effects of today's rapidly changing scene of corporate ownerships through mergers and the incessant trading of shares. Prospective investors are identified with the aid of firms that find information by searching through computer-based data banks. These names and associated profiles are made available for sale and eventually reach brokers, who will use them to offer shares for sale by phone. In this way, ownership rapidly fluctuates. Given these conditions, plus the risk of corporate takeovers that sometimes lead to massive layoffs, attitude-loyalty becomes hard to maintain.[18]

When engineering codes of ethics mention collegiality, they generally cite acts that constitute *dis*loyalty. The NSPE code,

---

[17] The concept of dependent virtues is developed by Michael Slote in *Goods and Virtues* (Oxford: Clarendon Press, 1983).

[18] Joanne B. Ciulla, *The Working Life* (New York: Times Books, 2000); Richard Sennett, *The Corrosion of Character: The Personal Consequences of Work in the New Capitalism* (New York: W. W. Norton, 1998).

for example, states that "Engineers shall not attempt to injure, maliciously or falsely, directly or indirectly, the professional reputation, prospects, practice or employment of other engineers. Engineers who believe others are guilty of unethical or illegal practice shall present such information to the proper authority for action."

These injunctions not to defame colleagues unjustly and not to condone unethical practice are important, but collegiality also has a more positive dimension. Craig Ihara suggests that "Collegiality is a kind of connectedness grounded in respect for professional expertise and in a commitment to the goals and values of the profession, and . . . as such, collegiality includes a disposition to support and cooperate with one's colleagues."[19] In other words, the central elements of collegiality are: (1) respect for colleagues, valuing their professional expertise and their devotion to the social goods promoted by the profession; (2) commitment, in the sense of sharing a devotion to the moral ideals inherent in one's profession, and (3) connectedness, or awareness of participating in cooperative projects based on shared commitments and mutual support. As such, collegiality is a virtue defining the teamwork essential for pursuing shared goods.

Managers and Engineers

Respect for authority is important in meeting organizational goals. Decisions must be made in situations where allowing everyone to exercise unrestrained individual discretion would create chaos. Moreover, clear lines of authority provide a means for identifying areas of personal responsibility and accountability.

The relevant kind of authority has been called *executive authority:* the corporate or institutional right given to a person to exercise power based on the resources of an organization.[20] It is distinguishable from *power* (or influence) in getting the job done. It is distinguishable, too, from *expert authority:* the possession of special knowledge, skill, or competence to perform some task or to give sound advice. Employees respect authority when they accept the guidance and obey the directives issued by the employer having to do with the areas of activity covered by the employer's institutional authority, assuming the directives are legal and do not violate norms of moral decency.

---

[19] Craig K. Ihara, "Collegiality as a Professional Virtue," in *Professional Ideals,* ed. Albert Flores (Belmont, CA: Wadsworth, 1988), 60.

[20] Joseph A. Pichler, "Power, Influence and Authority," in *Contemporary Management,* ed. Joseph W. McGuire (Englewood Cliffs, NJ: Prentice Hall, 1974), 428; Richard T. De George, *The Nature and Limits of Authority* (Lawrence, KS: University Press of Kansas, 1985).

Within this general framework of authority, however, there are wide variations in how engineers and managers relate to each other. At one extreme, there is the rigid, top-down control. At the other extreme, there is something more like how professors and administrators interact within universities. Michael Davis and his colleagues found that corporations fall into three categories.[21]

"Engineer-oriented companies" focus primarily on the quality of products. Engineers' judgments about safety and quality are given great weight, and they are overridden rarely, when considerations such as cost and scheduling became especially important. "Customer-oriented companies" make their priority the satisfaction of customers. In these companies safety considerations are also given high priority, but engineers are expected to be more assertive in speaking as advocates for safety, so that it received a fair hearing amidst managers' preoccupation with satisfying the needs of customers. Because of this sharper differentiation of managers' and engineers' points of view, communication problems tend to arise more frequently. Finally, "finance-oriented companies" make profit the primary focus.

Davis reports that in addition to having different roles and authority, managers and engineers typically have different attitudes and approaches. Managers tend to be more distanced from the technical details of jobs; they focus more on jobs in their entirety, from wider perspectives; and they are more focused on people than things.

Professional Rights

We turn now to respect for the rights of engineers and others. Engineers have several types of moral rights, which fall into the sometimes overlapping categories of human, employee, contractual, and professional rights. As *humans,* engineers have fundamental rights to live and freely pursue their legitimate interests, which implies, for example, rights not to be unfairly discriminated against in employment on the basis of sex, race, or age. As *employees,* engineers have special rights, including the right to receive one's salary in return for performing one's duties and the right to engage in the nonwork political activities of one's choosing without reprisal or coercion from employers. As *professionals,* engineers have special rights that arise from their professional role and the obligations it involves. We begin with professional rights.

---

[21] Michael Davis, *Thinking Like an Engineer* (New York: Oxford University Press, 1998), 130–31.

Three professional rights have special importance: (1) the basic right of professional conscience, (2) the right of conscientious refusal, and (3) the right of professional recognition.

Right of Professional Conscience. The right of professional conscience is the moral right to exercise professional judgment in pursuing professional responsibilities. Pursuing those responsibilities involves exercising both technical judgment and reasoned moral convictions. This right has limits, of course, and must be balanced against responsibilities to employers and colleagues of the sort discussed earlier.

If the duties of engineers were so clear that it was obvious to every sane person what was morally proper in every situation, there would be little point in speaking of conscience in specifying this basic right. Instead, we could simply say it is the right to do what everyone agrees it is obligatory for the professional engineer to do. But engineering, like other professions, calls for morally complex decisions. It requires autonomous moral judgment in trying to uncover the most morally reasonable courses of action, and the correct courses of action are not always obvious.

As with most moral rights, the basic professional right is an entitlement giving one the moral authority to act without interference from others. It is a "liberty right" that places an obligation on others not to interfere with its proper exercise. Yet occasionally, special resources may be required by engineers seeking to exercise the right of professional conscience in the course of meeting their professional obligations. For example, conducting an adequate safety inspection may require that special equipment be made available by employers. Or, more generally, to feel comfortable about making certain kinds of decisions on a project, the engineers involved need an ethical climate conducive to trust and support, which management may be obligated to help create and sustain. In this way the basic right is also in some respects a "positive right" placing on others an obligation to do more than merely not interfere.

There are two general ways to justify the basic right of professional conscience. One is to proceed piecemeal by reiterating the justifications given for the specific professional *duties*. Whatever justification there is for the specific duties will also provide justification for allowing engineers the *right* to pursue those duties. Fulfilling duties, in turn, requires the exercise of moral reflection and conscience, rather than rote application of simplistic rules. Hence the justification of each duty ultimately yields a justification of the right of conscience with respect to that duty.

The second way is to justify the right of professional conscience directly, which involves grounding it more directly in the ethical theories. Thus, duty ethics regards professional rights as implied by general duties to respect persons, and rule-utilitarianism would accent the public good of allowing engineers to pursue their professional duties. Rights ethics would justify the right of professional conscience by reference to the rights of the public not to be harmed and the right to be warned of dangers from the "social experiments" of technological innovation.

Right of Conscientious Refusal. The right of conscientious refusal is the right to refuse to engage in unethical behavior and to refuse to do so solely because one views it as unethical. This is a kind of second-order right. It arises because other rights to honor moral obligations within the authority-based relationships of employment sometimes come into conflict.

There are two situations to be considered: (1) where there is widely shared agreement in the profession as to whether an act is unethical and (2) where there is room for disagreement among reasonable people over whether an act is unethical.

It seems clear enough that engineers and other professionals have a moral right to refuse to participate in activities that are illegal and clearly unethical (for example, forging documents, altering test results, lying, giving or taking bribes, or padding payrolls). And coercing employees into acting by means of threats (to their jobs) plainly constitutes a violation of this right of theirs.

The troublesome cases concern situations where there is no shared agreement about whether a project or procedure is unethical. Do engineers have any rights to exercise their personal consciences in these more cloudy areas? Just as pro-life physicians and nurses have a right not to participate in abortions, engineers should be recognized as having a *limited* right to turn down assignments that violate their personal consciences in matters of great importance, such as threats to human life, even where there is room for moral disagreement among reasonable people about the situation in question. We emphasize the word *limited* because the right is contingent on the organization's ability to reassign the engineer to alternative projects without serious economic hardship to itself. The right of professional conscience does not extend to the right to be paid for not working.

Right of Recognition. Engineers have a right of professional recognition for their work and accomplishments. Part of this involves fair monetary remuneration, and part nonmonetary forms of recognition. The right to recognition, and especially fair remuneration, may seem to be purely a matter of self-interest rather than morality, but it is both. Without a fair remuneration, engineers cannot concentrate their energies where they properly belong—on carrying out the immediate duties of their jobs and on maintaining up-to-date skills through formal and informal continuing education. Their time will be taken up by money worries, or even by moonlighting to maintain a decent standard of living.

The right to reasonable remuneration is clear enough to serve as a moral basis for arguments against corporations that make excessive profits while engineers are paid below the pay scales of blue-collar workers. It can also serve as the basis for criticizing the unfairness of patent arrangements that fail to give more than nominal rewards to the creative engineers who make the discoveries leading to the patents. If a patent leads to millions of dollars of revenue for a company, it is unfair to give the discoverer no more than a nominal bonus and a thank-you letter.

But the right to professional recognition is not sufficiently precise to pinpoint just what a reasonable salary is or what a fair remuneration for patent discoveries is. Such detailed matters must be worked out cooperatively between employers and employees, for they depend on both the resources of a company and the bargaining position of engineers. Professional societies can be of help by providing general guidelines.

Employee Rights

Employee rights are any rights, moral or legal, that involve the status of being an employee. They overlap with some professional rights, of the sort just discussed, and they also include institutional rights created by organizational policies or employment agreements, such as the right to be paid the salary specified in one's contract. However, here we will focus on human rights that exist even if unrecognized by specific contract arrangements.

Many of these human rights are discussed more fully in *Freedom Inside the Organization* by David Ewing who, as editor of *The Harvard Business Review* was very much part of the business mainstream.[22] Ewing refers to employee rights as the "black hole in American rights." The Bill of Rights in the Constitution was written to apply to government, not to business. But when the Constitution was written, no one envisaged the giant corporations that have emerged in our century. Corporations wield enormous power politically and socially, often in multinational settings; they operate much as mini-governments, and they are

---

[22] David W. Ewing, *Freedom Inside the Organization* (New York: McGraw-Hill, 1977), 234–35.

often comparable in size to those governments the authors of the Constitution had in mind. For example, American Telephone & Telegraph in the 1970s employed twice the number of people that inhabited the largest of the original 13 colonies when the Constitution was written.

Ewing proposes that large corporations ought to recognize a basic set of employee rights. As examples we will discuss rights to privacy and to equal opportunity.

Privacy Right. The right to pursue outside activities can be thought of as a right to personal privacy in the sense that it means the right to have a private life off the job. In speaking of the right to privacy here, however, we mean the right to control the access to and the use of information about oneself. As with the right to outside activities, this right is limited in certain instances by employers' rights, but even then who among employers has access to confidential information is restricted. For example, the personnel division needs medical and life insurance information about employees, but immediate supervisors usually do not.

Consider a few examples of situations in which the functions of employers conflict with the right employees have to privacy:

1. Job applicants at the sales division of an electronics firm are required to take personality tests that include personal questions about alcohol use and sexual conduct.
2. A supervisor unlocks and searches the desk of an engineer who is away on vacation without the permission of that engineer. The supervisor suspects the engineer of having leaked information about company plans to a competitor and is searching for evidence to prove those suspicions.
3. A large manufacturer of expensive pocket computers has suffered substantial losses from employee theft. It is believed that more than one employee is involved. Without notifying employees, hidden surveillance cameras are installed.
4. A rubber products firm has successfully resisted various attempts by a union to organize its workers. It is always one step ahead of the union's strategies, in part because it monitors the phone calls of employees who are union sympathizers. It also pays selected employees bonuses in exchange for their attending union meetings and reporting on information gathered. It considered, but rejected as imprudent, the possibility of bugging the rest areas where employees were likely to discuss proposals made by union organizers.

We may disagree about which of these examples involve abuse of employer prerogatives. Yet the examples remind us of the importance of privacy and of how easily rights of privacy are abused. Employers should be viewed as having the same trust relationship with their employees concerning confidentiality that doctors have with their patients and lawyers have with their clients.[23]

Right to Equal Opportunity: Preventing Sexual Harassment. One definition of sexual harassment is: "the unwanted imposition of sexual requirements in the context of a relationship of unequal power."[24] It takes two main forms: quid pro quo and hostile work environment.

Quid pro quo includes cases where supervisors require sexual favors as a condition for some employment benefit (a job, promotion, or raise). It can take the form of a sexual threat (of harm) or sexual offer (of a benefit in return for a benefit). *Hostile work environment,* by contrast, is any sexually oriented aspect of the workplace that threatens employees' rights to equal opportunity. It includes unwanted sexual proposals, lewd remarks, sexual leering, posting nude photos, and inappropriate physical contact.

What is morally objectionable about sexual harassment? Sexual harassment is a particularly invidious form of sex discrimination, involving as it does not only the abuse of gender roles and authority relationships, but the abuse of sexual intimacy itself. Sexual harassment is a display of power and aggression through sexual means. Accordingly, it has appropriately been called "dominance eroticized."[25] Insofar as it involves coercion, sexual harassment constitutes an infringement of one's autonomy to make free decisions concerning one's body. But whether or not coercion and manipulation are used, it is an assault on the victim's dignity. In abusing sexuality, such harassment degrades people on the basis of a biological and social trait central to their sense of personhood.

Thus a duty ethicist would condemn it as violating the duty to treat people with respect, to treat them as having dignity and not merely as means to personal aggrandizement and gratification of one's sexual and power interests. A rights ethicist would see it as a serious violation of the human right to pursue one's work free from the pressures, fears, penalties, and insults that typically accompany sexual harassment. And a utilitarian would emphasize the impact it has on the victim's happiness and self-

---

[23] Mordechai Mironi, "The Confidentiality of Personnel Records," *Labor Law Journal* 25 (May 1974): 289.

[24] Catherine A. MacKinnon, *Sexual Harassment of Working Women* (New Haven, CT: Yale University Press, 1978), 1, 57–82.

[25] Ibid., 162.

fulfillment, and on women in general. This also applies to men who experience sexual harassment.

Right to Equal Opportunity: Nondiscrimination. Perhaps nothing is more demeaning than to be discounted because of one's sex, race, skin color, age, or political or religious outlook. These aspects of biological makeup and basic conviction lie at the heart of self-identity and self-respect. Such discrimination— that is, morally unjustified treatment of people on arbitrary or irrelevant grounds—is especially pernicious within the work environment, for work is itself fundamental to a person's self-image. Accordingly, human rights to fair and decent treatment at the workplace and in job training are vitally important.

Consider the following two examples:

1. An opening arises for a chemical plant manager. Normally such positions are filled by promotions from within the plant. The best qualified person in terms of training and years of experience is an African American engineer. Management believes, however, that the majority of workers in the plant would be disgruntled by the appointment of a nonwhite manager. They fear lessened employee cooperation and efficiency. They decide to promote and transfer a white engineer from another plant to fill the position.

2. A farm equipment manufacturer has been hit hard by lowered sales caused by a flagging produce economy. Layoffs are inevitable. During several clandestine management meetings, it is decided to use the occasion to "weed out" some of the engineers within 10 years of retirement to avoid payments of unvested pension funds.

These examples involve (immoral) discrimination. They also involve the violation of the Civil Rights Act of 1964 and the 1967 Age Discrimination in Employment Act, respectively.

Right to Equal Opportunity: Affirmative Action. Affirmative action, as the expression is usually defined, is giving a preference or advantage to a member of a group that in the past was denied equal treatment, in particular, women and minorities. The *weak form* of preferential treatment consists in hiring a woman or a member of a minority over an equally qualified white male. The *strong form,* by contrast, consists in giving preference to women or minorities over better-qualified white males.

Affirmative action began to be used during the 1960s. A major legal challenge to it came in 1978 in the *Regents of the University of California v. Bakke.* In that case, Allan Bakke, a white male engineer, was denied entrance to the medical school at the University of California, Davis (UC Davis), which reserved 16 of 100 openings for applicants who were either black, Latino, Asian, or American Indian. He sued, arguing that his credentials were superior to many of the minority students accepted. The Supreme Court ruled that the UC Davis admissions program was unconstitutional because it used explicit numerical quotas for minorities, which prevented person-to-person comparisons among all applicants. Yet the court also ruled that using race as one of many factors in comparing applicants is permissible, as long as quotas are avoided, and the intent is to ensure the goal of diversity among students—an important educational goal.

The same basic line of reasoning was reaffirmed in two Supreme Court rulings concerning the University of Michigan on June 23, 2003. In *Grutter v. Bollinger,* the Court approved of the University of Michigan's law school, which took race into account as one of many factors to achieve a diverse student body, ensuring a "critical mass" of minority students who could feel comfortable in expressing their viewpoints without being narrowly stereotyped (which tends to occur when there is only a token representation of minorities). In *Gratz v. Bollinger,* by contrast, the court ruled that the University of Michigan's undergraduate admissions program was unconstitutional. That program gave an automatic 20 points to members of minorities, out of the 100 needed for entrance (and out of a total possible 150 points). Such a rigid point system, the Court ruled, functioned too much like a quota system.

The rulings in both *Bakke* and *Grutter* were close: 5 to 4. Furthermore, in *Grutter* the Court made it clear that eventually, certainly within the next 25 years, it was expected that there would no longer be a need for affirmative action programs. Ironically, by 2003, many businesses and the military, which had in the 1960s opposed affirmative action, joined most educational institutions in desiring affirmative action policies as a way to achieve diversity for their own needs. Yet, even within education, there is no consensus on the issue. For example, in 1996 California voted (in Proposition 209) to forbid the use of race in granting admission to public universities, and that ruling still stands. In short, affirmative action remains a lively and contentious issue because of the important and clashing moral values at stake.

Can such preferences, either in the weak or strong form, ever be justified morally (as distinct from legally)? There are strong arguments on both sides of the issue.[26]

---

[26] Steven M. Cahn, ed., *The Affirmative Action Debate, 2nd ed.* (New York: Routledge, 2002); Carl Cohen and James P. Sterba, *Affirmative Action and Racial Preference: A Debate* (New York: Oxford University Press, 2003).

Arguments favoring preferential treatment take three main forms, which look to the past, present, and future. First, there is an argument based on compensatory justice: Past violations of rights must be compensated. Ideally such compensation should be given to those specific individuals who in the past were denied jobs. But the costs and practical difficulties of determining such discrimination on a case-by-case basis through the job-interviewing process permits giving preference on the basis of membership in a group that has been disadvantaged in the past. Second, sexism and racism still permeate our society today, and to counterbalance their insidious impact reverse preferential treatment is warranted to ensure equal opportunity for minorities and women. Third, reverse preferential treatment has many good consequences: integrating women and minorities into the economic and social mainstream (especially in male-dominated professions such as engineering), providing role models for minorities that build self-esteem, and strengthening diversity in ways that benefit both business and the wider community.

Arguments against reverse preferential treatment condemn it as reverse discrimination. It violates the rights to equal opportunity of white males and others who are now not given a fair chance to compete on the basis of their qualifications. Granted, past violations of rights may call for compensation, but only compensation to specific individuals who are wronged and only in ways that do not violate the rights of others who did not personally wrong minorities. It is also permissible to provide special funding for educational programs for economically disadvantaged children but not to use jobs as a compensatory device. Moreover, reverse preferential treatment has many negative effects: lowering economic productivity by using criteria other than qualifications in hiring, encouraging racism by generating intense resentment among white males and their families, encouraging traditional stereotypes that minorities and women cannot make it on their own without special help, and thereby adding to self-doubts of members of these groups.

Various attempts have been made to develop intermediate positions sensitive to all of the preceding arguments for and against strong preferential treatment. For example, one approach rejects blanket preferential treatment of special groups as inherently unjust, but it permits reverse preferential treatment within companies that can be shown to have a history of bias against minorities or women. Another approach is to permit weak reverse preferential treatment but to forbid strong forms.

Discussion Questions

1. Present and defend your view as to whether affirmative action is morally permissible and desirable in (a) admissions to engineering schools, (b) hiring and promoting within engineering corporations.

2. The majority of employers have adopted mandatory random drug testing on their employees, arguing that the enormous damage caused by the pervasive use of drugs in our society carries over into the workplace. Typically the tests involve taking urine or blood samples under close observation, thereby raising questions about personal privacy as well as privacy issues about drug use away from the workplace that is revealed by the tests. Present and defend your view concerning mandatory drug tests at the workplace.

   In your answer, take account of the argument that, except where safety is a clear and present danger (as in the work of pilots, police, and the military), such tests are unjustified.[27] Employers have a right to the level of performance for which they pay employees, a level typically specified in contracts and job descriptions. When a particular employee fails to meet that level of performance, then employers will take appropriate disciplinary action based on observable behavior. Either way, it is employee performance that is relevant in evaluating employees, not drug use per se.

3. A company advertises for an engineer to fill a management position. Among the employees the new manager is to supervise is a woman engineer, Ms. X, who was told by her former boss that she would soon be assigned tasks with increased responsibility. The prime candidate for the manager's position is Mr. Y, a recent immigrant from a country known for confining the roles for women. Ms. X was alerted by other women engineers to expect unchallenging, trivial assignments from a supervisor with Mr. Y's background. Is there anything she can and should do? Would it be ethical for her to try to forestall the appointment of Mr. Y?

4. Jim Serra, vice president of engineering, must decide who to recommend for a new director-level position that was formed by merging the product (regulatory) compliance group with the environmental testing group.[28] The top inside candidate is Diane Bryant, senior engineering group manager in charge of the environmental testing group. Bryant is 36, exceptionally intelligent and highly motivated, and a well-respected leader. She is also five months pregnant and is expected to take an eight-week maternity leave two months before the first customer ship deadline (six months away) for a new product. Bryant applies for the

---

[27] Joseph DesJardins and Ronald Duska, "Drug Testing in Employment," *Business & Professional Ethics Journal* 6 (1987): 3–21.

[28] This case is a summary of Cindee Mock and Andrea Bruno, "The Expectant Executive and the Endangered Promotion," *Harvard Business Review* (January–February 1994): 16–18.

job and in a discussion with Serra assures him that she will be available at all crucial stages of the project. Your colleague David Moss, who is vice president of product engineering, strongly urges you to find an outside person, insisting that there is no guarantee that Bryant will be available when needed. Much is at stake. A schedule delay could cost several million dollars in revenues lost to competitors. At the same time, offending Bryant could lead her and perhaps other valuable engineers whom she supervises to leave the company. What procedure would you recommend in reaching a solution?

5. In the past, engineering societies have generally portrayed participation by engineers in unions and collective bargaining in engineering as unprofessional and disloyal to employers. Critics reply that such generalized prohibitions reflect the excessive degree to which engineering is still dominated by corporations' interests. Discuss this issue with regard to the following case. What options might be pursued, and would they still involve "collective coercive action"?

Managers at a mining and refinery operation have consistently kept wages below industry-wide levels. They have also sacrificed worker safety to save costs by not installing special structural reinforcements in the mines, and they have made no effort to control excessive pollution of the work environment. As a result, the operation has reaped larger-than-average profits. Management has been approached both by individuals and by representatives of employee groups about raising wages and taking the steps necessary to ensure worker safety, but to no avail. A nonviolent strike is called, and the metallurgical engineers support it for reasons of worker safety and public health.

CHAPTER 7

Truth and Truthfulness

The Bay Area Rapid Transit system (BART) is a suburban rail system, constructed during the late 1960s and early 1970s, that links San Francisco with the cities across its bay. The opportunity to build a rail system from scratch, unfettered by old technology, was a challenge that excited many engineers and engineering firms. Yet among the engineers who worked on it were some who came to feel that too much "social experimentation" was going on without proper safeguards. Their zealous pursuit of the truth resulted in a classic case of whistle-blowing.[1]

Three engineers in particular, Holger Hjortsvang, Robert Bruder, and Max Blankenzee, identified dangers that were to be recognized by management only much later. They saw that the automatic train control was unsafely designed. Moreover, schedules for testing it and providing operator training prior to its public use were inadequate. Computer software problems continued to plague the system. Finally, there was insufficient monitoring of the work of the various contractors hired to design and construct the railroad. These inadequacies were to become the main causes of several early accidents.

The three engineers wrote a number of memos and voiced their concerns to their employers and colleagues. Their initial efforts were directed through organizational channels to both their immediate supervisors and the two next higher levels of management, but to no avail. They then took some controversial steps. Hjortsvang wrote an anonymous memo summarizing the problems and distributed copies of it to nearly all levels of management, including the project's general manager. Later, the three engineers contacted several members of BART's board of directors when their concerns were not taken seriously by lower

---

[1] Robert M. Anderson et al., *Divided Loyalties* (West Lafayette, Indiana: Purdue University Press, 1980).

levels of management. These acts constituted whistle-blowing within the organization.

One of the directors, Dan Helix, listened sympathetically to the engineers and agreed to contact top management while keeping their names confidential. But to the shock of the three engineers, Helix released copies of their unsigned memos and the consultant's report to the local newspapers. It would be the engineers, not Helix, who would be penalized for this act of whistle-blowing outside the organization.

Management immediately sought to locate the source of Helix's information. Fearing reprisals, the engineers at first lied to their supervisors and denied their involvement. At Helix's request the engineers later agreed to reveal themselves by going before the full board of directors to seek a remedy for the safety problems. On that occasion they were unable to convince the board of those problems. One week later they were given the option of resigning or being fired. The grounds given for the dismissal were insubordination, incompetence, lying to their superiors, causing staff disruptions, and failing to follow organizational procedures.

The dismissals were very damaging to the engineers. Robert Bruder could not find engineering work for eight months. He had to sell his house, go on welfare, and receive food stamps. Max Blankenzee was unable to find work for nearly five months, lost his house, and was separated from his wife for one and a half months. Holger Hjortsvang could not obtain full-time employment for 14 months, during which time he suffered from extreme nervousness and insomnia. Two years later the engineers sued BART for damages on the grounds of breach of contract, harming their future work prospects, and depriving them of their constitutional rights under the First and Fourteenth Amendments. A few days before the trial began, however, they were advised by their attorney that they could not win the case because they had lied to their employers during the episode. They settled out of court for $75,000 minus 40 percent for lawyers' fees. In the development of their case the engineers were assisted by an amicus curiae ("friend of the court") brief filed by the IEEE. This legal brief noted in their defense that it is part of each engineer's professional duty to promote the public welfare, as stated in IEEE's code of ethics. In 1978 IEEE presented each of them with its Award for Outstanding Service in the Public Interest for "courageously adhering to the letter and spirit of the IEEE code of ethics."

The case illustrates some of the moral complexities surrounding whistle-blowing in pursuing and revealing the truth about safety and other important moral matters. After discussing these complexities, we turn to a second area in which truth and truth-

fulness play an especially important role: academic integrity and
research.

No topic in engineering ethics is more controversial than whistle-
blowing. Is whistle-blowing morally permissible? Is it ever morally obligatory, or is it beyond the call of duty? To what extent, if
any, do engineers have a right to whistle-blow, and when is doing
so immoral and imprudent? When is whistle-blowing an act of
disloyalty to an organization? What procedures ought to be followed in blowing the whistle?

Whistle-Blowing: Definition

Whistle-blowing occurs when an employee or former employee
conveys information about a significant moral problem to someone in a position to take action on the problem, and does so
outside approved organizational channels (or against strong
pressure). The definition has four main parts.

1. *Disclosure:* Information is intentionally conveyed outside
   approved organizational (workplace) channels or in situations
   where the person conveying it is under pressure from supervisors or others not to do so.
2. *Topic:* The information concerns what the person believes is a
   significant moral problem for the organization (or an organization
   with which the company does business). Examples of significant
   problems are serious threats to public or employee safety and
   well-being, criminal behavior, unethical policies or practices, and
   injustices to workers within the organization.
3. *Agent:* The person disclosing the information is an employee or
   former employee, or someone else closely associated with the
   organization (as distinct, say, from a journalist reporting what the
   whistle-blower says).
4. *Recipient:* The information is conveyed to a person or organization that is in a position to act on the problem (as distinct, for
   example, to telling it to a family member or friend who is in no
   position to do anything).[2] The desired response or action might
   consist in remedying the problem or merely alerting affected
   parties.

Using this definition, we will speak of *external whistle-
blowing* when the information is passed outside the organization.

---

[2] We adopt the fourth condition from Marcia P. Miceli and Janet P. Near,
*Blowing the Whistle: The Organizational and Legal Implications for Companies
and Employees* (New York: Lexington Books, 1992), 15.

Internal whistle-blowing occurs when the information is conveyed to someone within the organization (but outside approved channels or against pressures to remain silent). The definition also allows us to distinguish between *open whistle-blowing,* in which individuals openly reveal their identity as they convey the information, and a*nonymous whistle-blowing,* which involves concealing one's identity. Some instances are partly open and partly anonymous, such as when individuals acknowledge their identities to a journalist but insist their names be withheld from anyone else.

Notice that the definition does not mention the motives involved in the whistle-blowing, and hence it avoids assumptions about whether those motives are good or bad. Nor does it assume that the whistle-blower is correct in believing there is a serious moral problem. In general, it leaves open the question of whether whistle-blowing is justified. BART is a case of both internal and external whistle-blowing that IEEE regarded as desirable and admirable. Let us contrast it with a case in which tragedy occurred because whistle-blowing was not engaged in.

In 1974 the first crash of a fully loaded DC-10 jumbo jet occurred over the suburbs of Paris; 346 people were killed, a record for a single-plane crash. It was known in advance that such a crash was bound to occur because of the jet's defective design.[3] The fuselage of the plane was developed by Convair, a subcontractor for McDonnell Douglas. Two years earlier, Convair's senior engineer directing the project, Dan Applegate, had written a memo to the vice president of the company itemizing the dangers that could result from the design. He accurately detailed several ways the cargo doors could burst open during flight, depressurize the cargo space, and thereby collapse the floor of the passenger cabin above. Because control lines ran along the cabin floor, this would mean a loss of full control over the plane. Applegate recommended redesigning the doors and strengthening the cabin floor. Without such changes, he stated, it was inevitable that some DC-10 cargo doors would open in midair, resulting in crashes.

In responding to this memo, top management at Convair disputed neither the technical facts cited by Applegate nor his predictions. Company officials maintained, however, that the possible financial liabilities Convair might incur prohibited them from passing on this information to McDonnell Douglas. These liabilities could be severe because the cost of redesign and the delay to make the necessary safety improvements would be very

---

[3] John H. Fielder and Douglas Birsch, eds., *The DC-10 Case* (Albany, NY: State University of New York Press, 1992); Paul Eddy, Elaine Potter, and Bruce Page, *Destination Disaster* (New York: Quadrangle, 1976).

high and would occur at a time when McDonnell Douglas would
be placed at a competitive disadvantage.

Moral Guidelines

Under what conditions are engineers justified in whistle-blowing?
This really involves two questions: When are they morally permitted, and when are they morally obligated, to do so? In our view, it
is permissible to whistle-blow when the following conditions have
been met.[4] Under these conditions there is also an obligation to
whistle-blow, although the obligation is prima facie and in some
situations can be overridden by other moral considerations.

1. The actual or potential harm reported is serious.
2. The harm has been adequately documented.
3. The concerns have been reported to immediate superiors.
4. After not getting satisfaction from immediate superiors, regular
   channels within the organization have been used to reach up to
   the highest levels of management.
5. There is reasonable hope that whistle-blowing can help prevent
   or remedy the harm.

These conditions are not always necessary for permissible and
obligatory whistle-blowing, however.[5] Condition 2 might not be
required in situations where cloaks of secrecy are imposed on
evidence that, if revealed, could supposedly aid commercial competitors or a nation's adversaries. In such cases it might be very
difficult to establish adequate documentation, and the whistle-
blowing would consist essentially of a request to the proper
authorities to carry out an external investigation or to request a
court to issue an order for the release of information.

Again, conditions 3 and 4 may be inappropriate in some situations, such as when one's supervisors are the main source of
the problem or when extreme urgency leaves insufficient time to
work through all regular organizational channels.

Finally, when whistle-blowing demands great sacrifices, one
cannot overlook that personal obligations to family, as well as
rights to pursue one's career, militate against whistle-blowing.
Where blowing the whistle openly could result not only in the loss
of one's job but also in being blacklisted within the profession,
the sacrifice may become supererogatory—more than is morally

---

[4] Adapted from Richard T. De George, "Ethical Responsibilities of Engineers
in Large Organizations: The Pinto Case," *Business and Professional Ethics
Journal* 1 (Fall 1981): 6.

[5] Gene G. James, "Whistle Blowing: Its Moral Justification," in *Business
Ethics,* 4th ed., ed. W. Michael Hoffman, Robert E. Frederick, and Mark S.
Schwartz (Boston: McGraw-Hill, 2001), 291–302.

obligatory. Engineers share responsibilities with many others for the products they help create. It seems unfair to demand that one individual bear the harsh penalties for picking up the moral slack for other irresponsible persons involved. Most important, the public also shares some responsibilities for technological ventures and hence for passing reasonable laws protecting responsible whistle-blowers. When those laws do not exist or are not enforced, the public has little basis for demanding that engineers risk their means of livelihood.[6]

Would Applegate have been justified had he decided to whistle-blow? As a loyal employee Applegate had a responsibility to follow company directives, at least reasonable ones. Perhaps he also had family responsibilities that made it important for him not to jeopardize his job. Yet as an engineer he was obligated to protect the safety of those who would use or be affected by the products he designed. Given the great public hazard involved, few would question whether it would be morally *permissible* for him to blow the whistle, either to the FAA or to the newspapers. Was he also morally *obligated* to blow the whistle? We leave this as a discussion question.

Not all whistle-blowing, of course, is admirable, obligatory, or even permissible. Certainly inaccurate whistle-blowing can cause unjustified harm to companies that unfairly receive bad publicity that hurts employees, stockholders, and sometimes the economy.[7] But is there a general presumption against whistle-blowing that at most is overridden in extreme situations? Taken together, loyalty, collegiality, and respect for authority do create a presumption against whistle-blowing, but it is a presumption that can be overridden. Loyalty, collegiality, and respect for authority are not excuses or justification for shielding irresponsible conduct. In addition to these corporate virtues, there are public-spirited virtues, especially respect for the public's safety.

Protecting Whistle-Blowers

Most whistle-blowers have suffered unhappy fates. In the words of one lawyer who defended a number of them: "Whistle-blowing is lonely, unrewarded, and fraught with peril. It entails a substantial risk of retaliation which is difficult and expensive to challenge."[8] Yet the vital service to the public provided by many whistle-blowers has led increasingly to public awareness

---

[6] Mike W. Martin, "Whistle-Blowing," in *Meaningful Work: Rethinking Professional Ethics* (New York: Oxford University Press, 2000), 138-50.
[7] Michael Davis, "Avoiding the Tragedy of Whistle-blowing," *Business and Professional Ethics Journal* 8 (1989): 3–19.
[8] Peter Raven-Hansen, "Dos and Don'ts for Whistle-Blowers: Planning for Trouble," *Technology Review* 82 (May 1980): 44.

of a need to protect them against retaliation by employers. In particular, whistle-blowers played a vital role in informing the public and investigators about recent corporate and government scandals, including whistle-blowers Sherron Watkins at Enron, Cynthia Cooper at WorldCom, and Coleen Rowley at the Federal Bureau of Investigation (FBI)—three women who appeared on the cover of *Time Magazine* as "Persons of the Year" for 2002.[9]

Government employees have won important protections. Various federal laws related to environmental protection and safety and the Civil Service Reform Act of 1978 protect them against reprisals for lawful disclosures of information believed to show "a violation of any law, rule, or regulation, mismanagement, a gross waste of funds, an abuse of authority, or a substantial and specific danger to public health and safety."[10] In the private sector, employees are covered by statutes forbidding firing or harassing of whistle-blowers who report to government regulatory agencies the violations of some 20 federal laws, including those covering coal mine safety, control of water and air pollution, disposal of toxic substances, and occupational safety and health. In a few instances, unions provide further protection.[11]

Laws, when they are carefully formulated and enforced, provide two types of benefits for the public, in addition to protecting the responsible whistle-blower: episodic and systemic. The *episodic* benefits help to prevent harm to the public in particular situations. The *systemic* benefits send a strong message to industry to act responsibly or be subject to public scrutiny once the whistle is blown.

Laws alone will usually not suffice, however. When officialdom is not ready to enforce existing laws—or introduce obviously necessary laws—engineering associations and employee groups need to act as watchdogs ready with advice and legal assistance. Successful examples are the Government Accountability Project (GAP) and some professional societies that report the names of corporations found to have taken unjust reprisals against whistle-blowers.

Common Sense Procedures

It is clear that a decision to whistle-blow is a serious matter that deserves careful reflection. There are several rules of practical

---

[9] See the several essays and the cover story for "Persons of the Year," in *Time Magazine* (December 30, 2002/January 6, 2003), 8, 30–60.

[10] Ibid., 42; Also see Stephen H. Unger, *Controlling Technology: Ethics and the Responsible Engineer,* 2nd ed. (New York: Holt, Rinehart and Winston, 1992), 179–81.

[11] James C. Petersen and Dan Farrell, *Whistle-Blowing* (Dubuque, IA: Kendall/Hunt, 1986), 20.

advice and common sense that should be heeded before taking this action.[12]

1. Except for extremely rare emergencies, always try working first through normal organizational channels. Get to know both the formal and informal (unwritten) rules for making appeals within the organization.
2. Be prompt in expressing objections. Waiting too long may create the appearance of plotting for your advantage and seeking to embarrass a supervisor.
3. Proceed in a tactful, low-key manner. Be considerate of the feelings of others involved. Always keep focused on the issues themselves, avoiding any personal criticisms that might create antagonism and deflect attention from solving those issues.
4. As much as possible, keep supervisors informed of your actions, both through informal discussion and formal memorandums.
5. Be accurate in your observations and claims, and keep formal records documenting relevant events.
6. Consult trusted colleagues for advice—avoid isolation.
7. Before going outside the organization, consult the ethics committee of your professional society.
8. Consult a lawyer concerning potential legal liabilities.

Beyond Whistle-Blowing

Sometimes whistle-blowing is a practical moral necessity. But generally it holds little promise as the best possible method for remedying problems and should be viewed as a last resort.

The obvious way to remove the need for internal whistle-blowing is for management to allow greater freedom and openness of communication within the organization. By making those channels more flexible and convenient, the need to violate them would be removed. But this means more than merely announcing formal "open-door" policies and appeals procedures that give direct access to higher levels of management. Those would be good first steps, and a further step would be the creation of an ombudsperson or an ethics review committee with genuine freedom to investigate complaints and make independent recommendations to top management. The crucial factor that must be involved in any structural change, however, is the creation of an atmosphere of positive affirmation of engineers' efforts to assert and defend their professional judgments in matters involving ethical considerations.

---

[12] Stephen H. Unger, "How to Be Ethical and Survive," *IEEE Spectrum* 16 (December 1979): 56–57.

What about external whistle-blowing? Much of it can also be avoided by the same sorts of intra-organizational modifications. Yet there will always remain troublesome cases where top management and engineers differ in their assessments of a situation even though both sides may be equally concerned to meet their professional obligations to safety. To date, the assumption has been that management has the final say in any such dispute. But our view is that engineers have a right to some further recourse in seeking to have their views heard, including confidential discussions with the ethics committees of their professional societies.

Finally, conscientious engineers sometimes find the best solution to be to resign and engage in protest, as in the following example. David Parnas, a computer scientist, lost his initial enthusiasm for the Strategic Defense Initiative (SDI), also known as the Star Wars project.[13] He resigned from an advisory panel on computing after only the first meeting of the panel. When agency officials would not seriously listen to his doubts about the feasibility of the project, he gradually succeeded through journal articles, open debates, and public lectures to convince the profession that Star Wars did not differ much from conventional antiballistic missile defense without overcoming earlier shortcomings. Indeed, the system's complexity made it practically impossible to write software as reliable as it ought to be in tight-trigger situations. For his efforts on behalf of the public interest, he was honored with the Norbert Wiener Award by the society of Computer Professionals for Social Responsibility (CPSR).

Discussion Questions

1. Present and defend your view as to whether, and in what respects, the BART engineers and BART management acted responsibly. In doing so, discuss alternative courses of action that either or both groups might have pursued.

2. According to Kenneth Kipnis, a professor of philosophy, Dan Applegate and his colleagues share the blame for the death of the passengers in the DC-10 crash. Kipnis contends that the engineers' overriding obligation was to obey the following principle: "Engineers shall not participate in projects that degrade ambient levels of public safety unless information concerning those degradations is made generally available."[14] Do you agree

---

[13] Carl Page, "Star Wars, Down but Not Out," *Newsletter of Computer Professionals for Social Responsibility* 14, no. 4 (Fall 1996).

[14] Kenneth Kipnis, "Engineers Who Kill: Professional Ethics and the Paramountcy of Public Safety," *Business and Professional Ethics Journal* 1 (1981): 82.

or disagree with Kipnis, and why? Was Applegate obligated to blow the whistle?

3. Present and defend your view as to whether in the following case the actions of Ms. Edgerton and her supervisor were morally permissible, obligatory, or admirable.

Virginia Edgerton was senior information scientist on a project for New York City's Criminal Justice Coordinating Council. The project was to develop a computer system for use by New York district attorneys in keeping track of data about court cases. It was to be added to another computer system, already in operation, which dispatched police cars in response to emergency calls. Ms. Edgerton, who had 13 years of data processing experience, judged that adding on the new system might result in overloading the existing system in such a way that the response time for dispatching emergency vehicles might increase. Because it might risk lives to test the system in operation, she recommended that a study be conducted ahead of time to estimate the likelihood of such overload.

She made this recommendation to her immediate supervisor, the project director, who refused to follow it. She then sought advice from the IEEE, of which she was a member. The Institute's Working Group on Ethics and Employment Practices referred her to the manager of systems programming at Columbia University's computer center, who verified that she was raising a legitimate issue.

Next she wrote a formal memo to her supervisor, again requesting the study. When her request was rejected, she sent a revised version of the memo to New York's Criminal Justice Steering Committee, a part of the organization for which she worked. In doing so she violated the project director's orders that all communications to the Steering Committee be approved by him in advance. The project director promptly fired her for insubordination. Later he stated: "It is . . . imperative that an employee who is in a highly professional capacity, and has the exposure that accompanies a position dealing with top level policy makers, follow expressly given orders and adhere to established policy."[15]

4. A controversial area of recent legislation allows whistle-blowers to collect money. Federal tax legislation, for example, pays informers a percentage of the money recovered from tax violators. And the 1986 False Claims Amendment Act allows 15 to 25 percent of the recovered money to go to whistle-blowers

---

[15] "Edgerton Case," Reports of the IEEE-CSIT Working Group on Ethics & Employment and the IEEE Member Conduct Committee in the matter of Virginia Edgerton's dismissal as information scientist of New York City, reprinted in *Technology and Society* 22 (June 1978): 3–10.

who report overcharging in federal government contracts to corporations. These sums can be substantial because lawsuits can
involve double and triple damages as well as fines. One study
revealed the following statistics: Overall, approximately 23 percent of these lawsuits succeed; 80 percent succeed if the federal
government joins the case as a plaintiff; 5 percent succeed when
the government does not join the case.[16] Discuss the possible
benefits and drawbacks of using this approach in engineering
and specifically concerning safety matters. Is the added incentive to whistle-blow worth the risk of encouraging self-interested
motives in whistle-blowing?

5. Do you see any special moral issues raised by anonymous
whistle-blowing?[17]

Truthfulness

**7.2 Honesty
and Research
Integrity**

The standard of truthfulness in engineering is very high, much
higher than in everyday life. It imposes what many consider an
absolute prohibition on deception, and in addition it establishes
a high ideal of seeking and speaking the truth. It also generates an array of issues concerning truthfulness in research and
academia.

In everyday life, the exact requirements of truthfulness are
often blurry and contested. Most of us are deeply offended when
others deceive (intentionally mislead) us, especially when they lie
(intentionally state what they know is false). At the same time,
we know of situations that require something less than complete
candor, for example to protect our privacy from the intrusive
questions of a stranger. Sissela Bok, however, contends that our
society has gone too far in creating a climate of dishonesty. She
acknowledges the need for occasional lies, for example, to protect
innocent lives, and for instances of withholding truths to protect
privacy rights. Yet she urges us all to embrace what she calls
the *principle of veracity*: There is a strong presumption against
lying and deception, although the presumption can occasionally be overridden by other pressing moral reasons in particular
contexts.[18]

Even Bok's principle of veracity is too weak to capture the
standard of honesty in engineering. Because so much is at stake
in terms of human safety, health, and well-being, engineers are
required and expected to seek and to speak the truth conscien-

---

[16] Peter Pae, "For Whistle-blowers, Virtue May Be the Only Reward," *Los
Angeles Times,* June 16, 2003, A-1 and A-14.
[17] Frederick A. Elliston, "Anonymous Whistle-Blowing," *Business and
Professional Ethics Journal* 1 (1982): 39–58.
[18] Sissela Bok, *Lying: Moral Choice in Public and Private Life* (New York:
Vintage Books, 1979), 32.

tiously and to avoid all acts of deception. To be sure, confidentiality requirements limit what can be divulged, but there is a stronger presumption against lying than even Bok's principle of veracity.

Two of the six Fundamental Canons in the National Society of Professional Engineers (NSPE) Code of Ethics concern honesty. Canon 3 requires engineers to "Issue public statements only in an objective and truthful manner," and Canon 5 requires them to "Avoid deceptive acts." We will refer to these requirements, taken together, as the *truthfulness responsibility:* Engineers must be objective and truthful and must not engage in deception. All other engineering codes set forth a statement of this responsibility.

The truthfulness responsibility enters often into the cases discussed by the NSPE in its *Opinions of the Board of Ethical Review*. Here are three examples that the board viewed as violating the NSPE Code of Ethics.[19]

1. An engineer who is an expert in hydrology and a key associate with a medium-size engineering consulting firm gives the firm her two-week notice, intending to change jobs. The senior engineer-manager at the consulting firm continues to distribute the firm's brochure, which lists her as an employee of the firm. (Case 90–4)

2. A city advertises a position for a city engineer/public works director, seeking to fill the position before the incumbent director retires to facilitate a smooth transition. The top candidate is selected after an extensive screening process, and on March 10 the engineer agrees to start April 10. By March 15 the engineer begins to express doubts about being able to start on April 10, and after negotiations the deadline is extended to April 24, based on the firm commitment by the engineer to start on that date. On April 23 the engineer says he has decided not to take the position. (Case 89–2)

3. An engineer working in an environmental engineering firm directs a field technician to sample the contents of storage drums on the premises of a client. The technician reports back that the drums most likely contain hazardous waste, and hence require removal according to state and federal regulations. Hoping to advance future business relationships with the client, the engineer merely tells the client the drums contain "questionable material" and recommends their removal, thereby giving the client greater leeway to dispose of the material inexpensively. (Case 92–6)

---

[19] National Society of Professional Engineers, *Opinions of the Board of Ethical Review*, vol. 7 (Alexandria, VA: NSPE, 1994); also see www.nspe.org/eh2-ctoc.htm (accessed Oct. 16, 2008).

As these examples suggest, the truthfulness responsibility applies widely and rules out all types of deception. Certainly it forbids lying, that is, stating what one knows to be false with the intention of misleading others. It also forbids intentional distortion and exaggeration, withholding relevant information (except for confidential information), claiming undeserved credit, and other misrepresentations designed to deceive. And it includes culpable failures to be objective, such as negligence in failing to investigate relevant information and allowing one's judgment to be corrupted.

Trustworthiness

Exactly why is truthfulness so important, especially within engineering but also in general? One answer centers on respect for autonomy. To deceive other persons is to undermine their autonomy, their ability to guide their own conduct. Deceit is a form of manipulation that undermines their ability to carry out their legitimate pursuits, based on available truths relevant to those pursuits. In particular situations, this can cause additional harm. Deceivers use other people as "mere means" to their own purposes, rather than respecting them as rational beings with desires and needs. This amounts to a kind of assault on a person's autonomy.[20]

Most moral theories defend truthfulness along these lines. Duty ethics, for example, provides a straightforward foundation for truthfulness as a form of respect for a person's autonomy. Rights ethics translates that idea into respect for a person's rights to exercise autonomy (or liberty). Rule-utilitarianism emphasizes the good consequences that flow from a rule requiring truthfulness. Virtue ethics affirms truthfulness as a fundamental virtue, and it underscores how honesty contributes to desirable forms of character for engineers, the internal good of the social practice of engineering, and the wider community in which that practice is embedded.

In addition, each of these ethical theories highlights additional wrongs in how deception harms others in engineering. Dishonest engineering causes financial losses, injuries, and death. Especially important, violating the truthfulness responsibility undermines trust. Honesty has two primary meanings: (1) truthfulness, which centers on meeting responsibilities about truth, and (2) trustworthiness, which centers on meeting responsibilities about trust. The meanings are interwoven because untruthfulness violates trust, and because violations of trust typically involve deception.

---

[20] Sissela Bok, *Lying: Moral Choice in Public and Private Life,* 19.

Engineering, like all professions, is based on exercising expertise within fiduciary (trust) relationships to provide safe and useful products. Untruthfulness and untrustworthiness undermine expertise by corrupting professional judgments and communications. They also undermine the trust of the public, employers, and others who must rely on engineers' expertise. Sound engineering is honest; dishonesty is bad engineering.

Academic Integrity: Students

Honesty as an engineer begins with honesty in studying to become an engineer. Studies of colleges and universities reveal alarming statistics about academic integrity. According to one study, among schools lacking a strong honor code, three out of four students admitted to having engaged in academic dishonesty at least once during their college career.[21] Among the schools with an honor code, one in two students made the same admission. Academic dishonesty among students takes several forms.[22]

Cheating: intentionally violating the rules of fair play in any academic exercise, for example, by using crib notes or copying from another student during a test.

Fabrication: intentionally falsifying or inventing information, for example, by faking the results of an experiment.

Plagiarism: intentionally or negligently submitting others' work as one's own, for example, by quoting the words of others without using quotation marks and citing the source.

Facilitating academic dishonesty: intentionally helping other students to engage in academic dishonesty, for example, by loaning them your work.

Misrepresentation: intentionally giving false information to an instructor, for example, by lying about why one missed a test.

Failure to contribute to a collaborative project: failing to do one's fair share on a joint project.

Sabotage: intentionally preventing others from doing their work, for example, by disrupting their lab experiment.

Theft: stealing, for example, stealing library books or other students' property.

Why do students engage in academic dishonesty?[23] Studies reveal a variety of motivations, including performance worries, such as fear of low grades and desires for higher grades; responses to external pressures, such as heavy workload, parental pres-

---

[21] D. L. McCabe and L. K. Trevino, "Academic Dishonesty: Honor Codes and Other Contextual Influences," *Journal of Higher Education* 64 (1993): 522–38.

[22] This is a slightly modified version of the categories assembled by Bernard E. Whitley Jr. and Patricia Keith-Spiegel, *Academic Dishonesty: An Educator's Guide* (Mahwah, NJ: Lawrence Erlbaum Associates, 2002), 17.

[23] Ibid., 23–27.

sure, or losing financial aid; the belief that professors are unfair, whether in demanding too much or in how they design tests and grades; the desire to help a friend; the belief that because other students are cheating it is all right for me to do the same; and the belief that plagiarism is not a big deal, and that it is a "victimless crime" in which no one really gets hurt. We should also ask why students do not cheat, that is, why they meet standards of academic integrity. Here, too, there are many motives: the conviction that dishonesty is wrong and unfair; the conviction that cheating undermines the meaning of achievement; self-respect; respect for the teacher; and fear of getting caught.

Explanation is one thing, justification is another. Is there any justification for cheating, or are proffered justifications simply rationalization—that is, biased and distorted reasoning? As authors, and like most educators, we take a firm stand. Academic dishonesty is a serious offense. It violates fair procedures. It harms other students who do not cheat by creating an undeserved advantage in earning grades. It is untruthful and deceives instructors and other members of an academic community. It violates trust—the trust of professors, other students, and the public who expect universities to maintain integrity. It undermines one's own integrity. And it renders dishonest and hollow any achievement or recognition based on the cheating.

Given the seriousness of academic dishonesty, and aware that we are all vulnerable to temptation, what can be done to foster academic integrity?[24] Researchers make several recommendations. Universities, as organizations, need to create and maintain a culture of honesty. Honor codes, which set forth firm standards and require students and faculty to report that cheating is going on, make a dramatic difference, even though they are not enough. Especially important, universities must support professors and students who follow university policies in reporting cheating, refusing to bow to the current market mentality in higher education that is more concerned about losing a paying "customer" than about ensuring academic integrity.

In addition, professors need to maintain a climate of respect, fairness, and concern for students. Course requirements and restrictions need to be explained clearly and then implemented. Tests and assignments need to be reasonable in terms of matching the material studied in class, and helpful feedback should be provided as the course progresses. Opportunities to cheat should be minimized. Firm and enforced disciplinary procedures are essential. Just as the Internet has made cheating easier, so has

---

[24] Ibid., 43–155. See also Wilfried Decoo, *Crisis on Campus: Confronting Academic Misconduct* (Cambridge, MA: MIT Press, 2002), 44–45, 49–50.

detecting plagiarism been made easier using new Web services. And, we might add, teaching about academic integrity can be a valuable way to integrate an ethics component into courses.

Research Integrity

Truthfulness takes on heightened importance in research because research aims at discovering, expressing, and promulgating truth. Research in engineering takes place in many settings, including universities, government labs, and corporations. The exact moral requirements vary somewhat, according to the applicable guidelines and regulations, but the truthfulness responsibility applies in all settings. Moreover, research ethics has many facets, several of which we discuss: Defining research integrity and misconduct, conducting and reporting experiments, protecting research subjects, giving and claiming credit, and reporting misconduct.

Integrity in research is about promoting excellence (high quality) in pursuing truth, and this positive emphasis on excellence should be kept paramount in thinking about honesty in research. An emphasis on quality and excellence in research broadens research ethics to include much more than avoiding fraud.[25] Research should be guided by what Richard Feynman calls a kind of utter honesty—a kind of leaning over backwards. For example, if you're doing an experiment, you should report everything that you think might make it invalid—not only what you think is right about it: other causes that could possibly explain your results; and things you thought of that you've eliminated by some other experiment, and how they worked—to make sure the other fellow can tell they have been eliminated.[26]

Positive ideals and requirements of research ethics should be borne in mind as we attend to the concerns about research misconduct that rivet public attention. Indeed, even misconduct in research is given both wider and narrower definitions, developed in specific contexts and for different purposes. For example, if the purpose is to assure high-quality research, in all its dimensions, a wider definition might be adopted. Wide definitions typically emphasize honesty in conducting and reporting experiments, while also including theft, other misuses of research funds, and sexual harassment among researchers.

If instead the purpose is to punish wrongdoers, a narrow and legalistic definition is likely to be favored.[27] Narrow definitions

---

[25] Rosemary Chalk, "Integrity in Science: Moving into the New Millennium," *Science and Engineering Ethics* 5, no. 2 (1999): 181.

[26] Richard P. Feynman, *Surely You're Joking, Mr. Feynman!* (New York: Bantam, 1986), 311.

[27] David H. Guston, "Changing Explanatory Frameworks in the U.S. Government's Attempt to Define Research Misconduct," *Science and Engineering Ethics* 5, no. 2 (1999): 137–54.

tend to be favored by universities, corporations, and other groups whose members are liable to punishment for offenses, whereas government agencies have pushed for broader definitions. For example, the National Science Foundation (NSF) defines misconduct in science and engineering as: "fabrication, falsification, plagiarism, or other serious deviation from accepted practices in proposing, carrying out, or reporting results from activities funded by NSF; or retaliation of any kind against a person who reported or provided information about suspected or alleged misconduct and who has not acted in bad faith."[28] The clause "or other serious deviation from accepted practice" is controversial because it is vague and wide in scope, thereby causing understandable anxiety in the research community. It becomes less controversial when the word "intentional" is inserted before "deviation."

Historically, the clear-cut instances of scientific misconduct are intentional violations, as the nineteenth-century mathematician Charles Babbage emphasized. Babbage distinguished four types of deception and fraud in research.[29] *Forging* is deception intended to establish one's reputation as a researcher, whereas *hoaxing* is deception intended to last only for a while and then to be uncovered or disclosed, typically to ridicule those who were taken in by it. *Trimming* is selectively omitting bits of outlying data—results that depart furthest from the mean. His most famous category was *cooking*, a term still used today to refer to all kinds of selective reporting of results, falsifying of data, and massaging data in the direction that supports the result one prefers.

Although Babbage's emphasis on intentional misrepresentation is the most common way of defining research fraud, intent is sometimes difficult to prove. Moreover, what about *gross negligence*, in which a researcher unintentionally, but culpably, fails to meet the minimum standards for conducting and reporting research, and other forms of extreme incompetence? Most negligence results from lack of due care in setting up an experiment, for example, by failing to establish a reliable control group or failing to properly monitor an experiment as it unfolds. Negligence can also result from biases and self-deception.

Bias and Self-Deception

At a hastily called news conference on March 23, 1989, the president of the University of Utah, Chase Peterson, made a stunning

---

[28] See Caroline Whitbeck, *Ethics in Engineering Practice and Research* (New York: Cambridge University Press, 1998), 201.

[29] Charles Babbage, "Reflections on the Decline of Science in England and on Some of Its Causes," in M. C. Kelly, ed., *The Works of Charles Babbage,* vol. 7 (London: Pickering, 1989), 90–93.

announcement.[30] A new and potentially limitless source of energy
had been discovered by Stanley Pons, the chair of the university's
chemistry department, and Martin Fleischmann, a Southampton
University professor who collaborated with Pons. Soon dubbed
"cold fusion," the experiment outlined by Pons and Fleischmann
was extraordinarily simple. It consisted of an electrochemical cell
in which two electrodes, including one made of palladium, were
immersed in a liquid containing the hydrogen isotope deuterium.
According to Pons and Fleischmann, applying an electric current
forced deuterium to concentrate in the palladium in a manner
that caused hydrogen nuclei to fuse, thereby producing excess
heat, radiation, and radiation byproducts such as tritium.

The Utah announcement generated frenetic research around
the world, involving hundreds of researchers and tens of millions of dollars. Some researchers thought they confirmed the
Pons-Fleischmann results, but there were also quick reversals,
as hastily interpreted data were reexamined. The most careful
experiments failed completely to replicate the results of the Pons-
Fleischmann experiment. There is now a consensus in the scientific community that cold fusion does not occur. There is also a
consensus that the cold fusion episode is a cautionary tale about
how bias and self-deception, bolstered by external pressures, can
undermine sound research.

Pons and Fleischmann were well-respected electrochem-
ists. They had made mistakes, and Fleischmann especially was
known for his daring hypotheses that sometimes failed and other
times succeeded dramatically. But mistakes and creative daring
are integral to research. What was objectionable was their highly
unorthodox step of announcing the results of research that had
not yet been published in peer-reviewed journals. If they had simply published their results and allowed other researchers to con-
firm or refute their conclusions, we would be dealing with science
as usual. Instead, they became so caught up in the excitement
of extraordinary achievement, including the prospect of Nobel
prizes, that they allowed their judgment to be distorted and their
work to become careless.

Their failing lies somewhere between deliberate deception
(fraud) and unintentional error (simple sloppiness), in the
domain of *self-deception*. One form of self-deception is motivated
irrationality—that is, unreasonable belief that is motivated
by allowing one's judgment to be biased by wishes, hopes, self-

---

[30] John R. Huizenga, *Cold Fusion: The Scientific Fiasco of the Century* (New
York: Oxford University Press, 1993).

esteem, and fears.[31] The word *allowing* implies negligence, that is, the failure to take sufficient care to prevent biases from distorting one's thinking and observations. Another form of self-deception is more purposeful evasion. For example, researchers suspect an unpleasant reality, perhaps sensing that the data are going against what they want to believe. Then, instead of confronting the data honestly, they purposefully disregard the evidence or downplay its implications. The purpose and intention involved is typically unconscious or less than fully conscious. The truthfulness responsibility requires trying to overcome both forms of self-deception.

Institutions of higher education are under increasing economic pressures that can threaten professional judgment. Financially strapped universities now eagerly seek commercial ties, and corporations seek the expertise and prestige of university researchers. This combination can and often does lead to creative partnerships, but it also poses risks, as the "commercialization of research" in universities threatens objective judgment by engineers and scientists in several areas: secrecy, conflicts of interest, and attempts to manipulate research results.[32]

Protecting Research Subjects

Research in engineering sometimes involves experimental subjects and also (nonhuman) animals, especially when it overlaps with biomedical research. The standards for experimenters are extensive and detailed, and we will highlight only a few of them pertaining to human subjects.[33]

Experiments on humans are permissible only after obtaining the voluntary consent of human subjects. That means giving to experimental subjects (or their surrogate decision makers) all information about the risks, possible benefits, alternatives, exact procedures involved, and all other information a reasonable person would want to know before participating in an experiment. In addition, there must be no coercion, threats, or undue pressure. And the individual must have the capacity to make a reasonable decision about whether to participate.

Special safeguards are taken when experimental subjects other than competent adults are the research subjects. When children

---

[31] Mike W. Martin, *Self-Deception and Morality* (Lawrence, KS: University Press of Kansas, 1986); Herbert Fingarette, *Self-Deception* (Berkeley: University of California Press, 2000).

[32] Derek Bok, *Universities in the Marketplace: The Commercialization of Higher Education* (Princeton: Princeton University Press, 2003), especially pages 57–78, 139–56.

[33] The National Research Council's Committee on Science, Engineering and Public Policy, *On Being a Scientist*, 2nd ed. (Washington, DC: National Academy Press, 1992).

participate in experiments, an appropriate surrogate decision maker, usually the parents, must give voluntary informed consent, and usually it is required that the child can reasonably be expected to benefit from the procedure. Experimentation on institutionalized persons, for example in prisons or mental institutions, is either forbidden or requires especially high standards. That is because of the inherently coercive nature of institutions that control all aspects of a person's life.

The Nuremberg Code, written immediately after World War II, is the most important historical document requiring informed consent in research. It was developed in light of the Nazi horrors, and it has been flagrantly violated under authoritarian regimes. In addition, the United States has occasionally violated informed consent. For example, during World War II, the U.S. government conducted biological, chemical, and nuclear experiments on unsuspecting individuals.[34]

Giving and Claiming Credit

Often there are pressures on researchers to varnish the truth when competing for professional recognition, not only because it brings ego gratification but also because it might involve winning jobs, promotions, and income. Outright fraud of the following types also occurs.

Plagiarism, as defined earlier, is intentionally or negligently submitting others' work as one's own. In research, it is claiming credit for someone else's ideas or work without acknowledging it, in contexts where one is morally required to acknowledge it. The latter clause, about what is morally required, is important. In a novel, where footnoting is not customary, a brief quotation without quotation marks and a reference might be an acceptable "literary allusion," but in an essay by a student or professor it would be considered theft.

Failure to give credit occurs in many different settings within engineering, and the NSPE Board of Ethical Review frequently comments on them. In Case 92–1, for example, a city hires an engineer to design a bridge, and the engineer in turn subcontracts some key design work to a second engineer.[35] Months after the bridge is completed, the first engineer submits the design to a national design competition where it wins an award, but he fails to credit the work of the second engineer. As we might expect, the board ruled that such conduct violates the truthfulness rule.

---

[34] Jonathan D. Moreno, *Undue Risk: Secret State Experiments on Humans* (New York: Routledge, 2001).

[35] National Society of Professional Engineers, *Opinions of the Board of Ethical Review*, vol. VII (Alexandria, VA: National Society of Professional Engineers, 1994), 61.

Misrepresenting credentials is a second type of deception. Occasionally researchers forge credentials, creating widely publicized scandals. Fabrications about articles and credentials are relatively easy to uncover. Misrepresentations of credentials can take more subtle forms, however. One of the earliest cases discussed by the NSPE Board of Ethical Review (Case 79–5) was about an engineer who received a doctoral degree from a "diploma mill" organization that required no attendance or study at its facilities. The engineer then listed the degree on his professional correspondence and brochures. The NSPE board reasoned that listing a doctoral degree, especially without listing where it is from, is widely understood to convey that it constitutes an earned doctorate, and that hence the engineer was indeed using unprofessional deception.

Misleading listing of authorship, whether of articles or other documents, is another area where subtle deception occurs. Obviously it is unethical to omit a coauthor who makes a significant contribution to the research. But the order of authors in many disciplines, including engineering, is also usually understood to convey information about the relative contributions of the authors, with the earlier listing indicating greater contributions. To be sure, customs vary—and on this topic respect for customs is important to ensure truthfulness. In some disciplines, the listing order is not considered important, for example, in philosophy where alphabetical listing is common, or in some sciences where there can be dozens of coauthors. But in an engineering paper written by several coauthors, the order of names is significant. Hence, in the United States, it would be unethical for a dissertation supervisor to list his or her name first, when in fact the doctoral student did the primary research for the article.

Finally, there is a growing consensus that researchers have a responsibility to report misconduct by other researchers when the misconduct is serious and when they are in a position to document it. Yet typically there are strong pressures—from supervisors, colleagues, and others—not to report misconduct, and hence most instances fall into the category of whistle-blowing. Measures to protect individuals who responsibly report research misconduct are being implemented at research facilities, and, as noted earlier, the concept of research misconduct now applies to punitive measures taken against these individuals. More needs to be done, however, and there is still a stigma against "turning in" a colleague.[36]

---

[36] See Philip J. Hilts, "The Science Mob: The David Baltimore Case—And Its Lessons," in *Research Ethics: A Reader,* ed. Deni Elliott and Judy E. Stern (Hanover, NH: University Press of New England, 1997), 43–51.

Discussion Questions

1. With regard to each of the three NSPE examples described earlier under "truthfulness," discuss exactly what is at stake in whether the truthfulness responsibility is met. In doing so, identify the relevant rights, duties, and good and bad consequences involved.

2. Robert is a third-year engineering student who has been placed on probation for a low grade point average, even though he knows he is doing the best work he can. A concerned friend offers to help him by sitting next to him and "sharing" his answers during the next exam. Robert has never cheated on an exam before, but this time he is desperate. What should he do?

3. A student gives Elaine a copy of a professor's midterm exam from last year. (a) Is it all right for her to accept the exam, without asking any questions? (b) Elaine decides to ask about how the exam was obtained. She learns that the professor had required all copies of the exam sheet to be returned but had inadvertently missed this copy, which a student then circulated to selected other students. She decides to decline the exam, but does she have any additional responsibilities?

4. Kermit Vandivier had worked at B. F. Goodrich for five years, first in instrumentation and later as a data analyst and technical writer. In 1968 he was assigned to write a report on the performance of the Goodrich wheels and brakes commissioned by the Air Force for its new A7-D light attack aircraft. According to his account, he became aware of the design's limitations and of serious irregularities in the qualification tests.[37] The brake failed to meet Air Force specifications. After pointing out these problems, however, he was given a direct order to stop complaining and write a report that would show the brake qualified. He was led to believe that several layers of management were behind this demand and would accept whatever distortions might be needed because their engineering judgment assured them the brake was acceptable.

   Vandivier then drafted a 200-page report with dozens of falsifications and misrepresentations. Yet, he refused to sign it. Later he gave as excuses for his complicity the facts that he was 42 years old with a wife and six children. He had recently bought a home and felt financially unable to change jobs. He felt certain that he would have been fired if he had refused to participate in writing the report.

---

[37] Kermit Vandivier, "Engineers, Ethics and Economics," in *Conference on Engineering Ethics* (New York: American Society of Civil Engineers, 1975), 20–34.

a. Assuming for the moment that Vandivier's account of the events is accurate, present and defend your view as to whether Vandivier was justified in writing the report or not. Which moral considerations would you cite in defending your view?

b. Vandivier was a technical writer, not an engineer, to whom the truthfulness responsibility (as stated in an engineering code of ethics) applies. Does that matter morally? That is, would you answer section **a** of this question in the same manner, whether an engineer or technical writer were involved? Does the applicability of a code of ethics alter the ethics of the situation?

c. Vandivier's account of the events has been challenged. After consulting the record of congressional hearings about this case, John Fielder concluded that Vandivier's "claims that the brake was improperly tested and the report falsified are well-supported and convincing, but he overstates the magnitude of the brake's defects and, consequently, [exaggerates] the danger to the [test] pilot."[38] Comment on the difficulties in achieving complete honesty when Vandivier and other participants in such instances tell their side of the story. Also comment on the limitations and possible harm (such as to companies' reputations) in relying solely on the testimony of one participant.

5. The use of Agent Orange defoliants in Vietnam has only recently been officially recognized by the United States as a health hazard as former U.S. soldiers began to show symptoms of ill effects, long after scientists warned of its effects on farmers and their animals in the war zones of Vietnam. When, if ever, does a war justify exposing soldiers and noncombatants to substances that can affect humans in ways that can have long term effects?

---

[38] John Fielder, "Give Goodrich a Break," *Business and Professional Ethics Journal* 7 (1988): 3–25.

CHAPTER 8

Computer Ethics

In November of 2007, the United Kingdom's tax agency reported a major security breach to the police. Tax information, including intimate personal knowledge and bank account numbers, for some 25 million British citizens—almost half the country's population—had been loaded onto two disks and sent by mail courier to an audit office. To the astonishment of all, the disks had gone missing. Worse, the information on them was unencrypted. The British populace was thrown into a state of financial terror. The competency of the five-month-old government of Prime Minister Gordon Brown was brought into question. The case illustrates the scale of concerns about privacy and security in the age of computers, and also the unpredictable interaction of sophisticated technological structures and human fallibility.

Computers have become the technological backbone of society. Their degree of complexity, range of applications, and sheer numbers continue to increase. Through telecommunication networks they span the globe. Yet electronic computers are still only a few decades old, and it is difficult to foresee all the moral issues that will eventually surround them. The present state of computers is sometimes compared to that of the automobile in the early part of this century. At that time the impact of cars on work and leisure patterns, pollution, energy consumption, and sexual mores was largely unimagined. If anything, it is more difficult to envisage the eventual impact of computers because they are not limited to any one primary area of use as is a car's function in transportation.

It is already clear, however, that computers raise a host of difficult moral issues, many of them connected with basic moral concerns such as free speech, privacy, respect for property, informed consent, and harm.[1] To evaluate and deal with these issues, a new area of applied ethics called computer ethics has sprung up. Computer ethics has special importance for the new groups of professionals emerging with computer technology, for example, designers of computers, programmers, systems analysts, and operators. To the extent that engineers design, manufacture, and apply computers, computer ethics is a branch of engineering ethics. But the many professionals who use and control computers share the responsibility for their applications.

Some of the issues in computer ethics concern shifts in power relationships resulting from the new capacities of computers. Other issues concern property, and still others are about invasions of privacy. All these issues may involve "computer abuse": unethical or illegal conduct in which computers play a central role (whether as instruments or objects).

The Internet and Free Speech

The Internet has magnified all issues in computer ethics. The most powerful communication technology ever developed, and a technology used daily by hundreds of millions of people, the Internet gained widespread use only during the 1990s. Its modest beginning, or forerunner, came from a simple idea of J. C. R. Licklider.[2]

Licklider was a psychologist who had wide interests in the newly emerging computer technology. In 1960 he conceived of a human-computer symbiosis in which the powers of humans and computers were mutually enhancing.[3] The breadth of his vision, together with his administrative skills, led to his appointment a few years later as the director of the Advanced Research Projects Agency (ARPA) of the U.S. Department of Defense. He quickly saw that the variety of computer-involved military projects was becoming a Tower of Babel, and he wrote a revolutionary memo

---

[1] For an overview of issues affecting business, see Richard T. De George, "Ethical Issues in Information Technology," in *The Blackwell Guide to Business Ethics,* ed. Norman E. Bowie (Malden, MA: Blackwell, 2002), 267–88. There are now many useful anthologies on computer ethics, including Richard A. Spinello and Herman T. Tavani, eds., *Readings in CyberEthics*, 2nd ed. (Boston: Jones and Bartlett, 2004), and Terrell Ward Bynum and Simon Rogerson, eds., Computer Ethics and Professional Responsibility (Malden, MA: Blackwell Publishing, 2004).

[2] M. Mitchell Waldrop, *The Dream Machine: J. C. R. Licklider and the Revolution that Made Computing Personal* (New York: Penguin, 2001); and *John Naughton, A Brief History of the Future: The Origins of the Internet* (London: Weidenfeld, 1999).

[3] "Man-Computer Symbiosis," *IRE Transactions on Human Factors in Electronics,* vol. HFE-1 (March 1960): 4–11; reprinted in *In Memorium: J. C. R. Licklider, 1915–1990,* ed. Robert W. Taylor (Palo Alto, CA: Digital Systems Research Center Reports, 1990).

calling for a move toward a unified communication system. In 1969, ARPA funded projects in universities and corporations that created an ARPA network, or ARPANET. In the 1980s, some universities developed their own communications networks, and their eventual merging with ARPANET became the Internet, which is now a global network of networks, initially using the infrastructure of the telephone system and now carried by many telecommunication systems by wire, fiber, or wireless systems. The World Wide Web (Web), which is a service run on the Internet, emerged from the Hypertext Markup Language and transfer protocol developed at the European particle physics lab and is used in a multimedia format of text, pictures, sound, and video. During the early 1990s, the Web was opened to business, e-mail, and other uses that continue to expand.

It is now clear to all that the Internet provides a wellspring of new ways to be in contact with other people and with sources of information. It has also created greater convenience in ordering consumer items, paying bills, and trading stocks and bonds. Like other major "social experiments," it also has raised a host of new issues. One set of issues centers on free speech, including control of obscene forms of pornography, hate speech, spam (unwanted commercial speech), and libel.

In a wide sense, pornography is sexually explicit material intended primarily for sexual purposes (as distinct, say, from medical education). *Obscene* pornography is pornography that is immoral or illegal in many countries, and is not protected in the United States by the First Amendment rights to free speech. U.S. laws define obscenity as sexually explicit materials that appeal to sexual interests, lack serious literary, artistic, scientific, or other value, and are offensive to reasonable persons as judged by a community's standards. Needless to say, there is considerable disagreement about what this means, and the definition is relative to communities that might have differing standards. At the same time, there is wide agreement that child pornography and extremely violent and degrading portrayals of women are obscene, and most local communities have attempted to control them. The Internet has made such control extremely difficult, as images and texts can be transmitted easily from international sources to a child's home computer. There are now hundreds of thousands of pornographic Web sites, with hundreds more created each day, many of which contain obscene material.

Hate speech, unlike obscenity, is not forbidden constitutionally. Not surprisingly, then, the Internet has become a powerful resource for racist and anti-Semitic groups to spread their messages. Those messages were heard, for example, by Eric Harris and Dylan Klebold, who massacred their fellow students at

Columbine High School in 1999. And there is no question that this most powerful medium makes it much easier for hate groups to organize and expand.

Two types of control of pornography and hate speech have been attempted: top-down control by governments, and bottom-up controls by individuals and groups in the marketplace.[4] Top-down controls have been attempted by both Democrats and Republicans. For example, Congress passed the Communications Decency Act, signed by President Clinton in 1996, which forbade transmitting indecent and patently offensive material to minors. A year later the Supreme Court declared the act unconstitutional. In contrast, a 2001 federal statute, the Children's Internet Protection Act, required libraries receiving federal funds to use filters to block pornographic material from library computers used by the public. In 2003 the statute was upheld as constitutional by the Supreme Court in *United States v. American Library Association.* Legislatures and courts continue to seek reasonable balance between protecting free speech and advancing other important values.

Parents who purchase blocking or filtering software exemplify bottom-up controls. If those controls are extended from homes to schools and other public settings, according to Richard Spinello, certain procedures should be followed.[5] The controls should be voluntary, in the sense that the relevant constituencies are allowed full participation in the process. Web sites that provide rating services for screening material should openly acknowledge the criteria they use. They should avoid hidden political agendas. (CyberSitter, for example, presented itself as blocking child pornography but then also blocked access to the National Organization for Women.) And the level of blocking should be low-level, rather than at unnecessarily blanket levels.

Some enthusiasts argue that controlling free speech on the Internet not only will prove unfeasible on any large scale, but also that such uncontrollability is good. The Internet could be the ultimate defender of principles of freedom, equality, and opportunity. It could gradually undermine the power of tyrants who have blocked democratic freedoms in their country, if unfettered communication channels can be maintained.

Power Relationships

Computers and the Internet dramatically increase the ability of centralized bureaucracies to manage enormous quantities of data,

---

[4] Richard Spinello, *Cyberethics: Morality and Law in Cyberspace* (Boston: Jones and Bartlett Publishers, 2000), 35–42.
[5] Ibid., 54.

involving multiple variables, and at astonishing speed. During the 1960s and 1970s social critics became alarmed at the prospect that computers would concentrate power in a few centralized bureaucracies of big government and big business, thereby eroding democratic systems by moving toward totalitarianism.

These fears were not unwarranted, but they have lessened because of recent developments in computer technology. In the early stages of computer development there were two good reasons for believing that computers would inevitably tend to centralize power.[6] Early large computers were many times cheaper to use when dealing with large tasks than were the many smaller computers it would have taken to perform similar tasks. Thus it seemed that economics would favor a few large and centrally located computers, suggesting a concentration of power in a few hands. Moreover, the large early computer systems could only be used by people geographically close to them, again implying that relatively few people would have access to them.

The development and proliferation of microcomputers changed all this. Small computers became increasingly powerful and economically competitive with larger models. Furthermore, remote access and time-sharing allowed computer users in distant locations to share the resources of large computer systems. These changes opened new possibilities for decentralized computer power. More recently, the purpose of linking computers has not been so much for the purpose of reaching machines with greater number-crunching capabilities as for the opportunities to exchange information. The physical links that make this possible, and the data processing technology that is associated therewith, make up the Internet.

Once, it was feared that computers would give the federal government far greater power to control nationally funded systems, such as the welfare and medical systems, lessening control by local and state governments. But in fact, data systems have turned out to be two-way, allowing both small governments and individuals to have much greater access to information resources amassed at the federal level.

Computers are powerful tools that do not by themselves generate power shifts. They contribute to greater centralization or decentralization insofar as human decision makers direct them. This is not to say that computers are entirely value-neutral. It is to say that moral issues about power relationships tend to be nuanced and contextual. A few examples follow.

---

[6] Herbert A. Simon, "The Consequences of Computers for Centralization and Decentralization," in *The Computer Age: A Twenty-Year View,* ed. Michael L. Dertouzos and Joel Moses (Cambridge, MA: MIT Press, 1979), 212–28.

Job Elimination. Computers have led and will continue to lead to the elimination of some jobs. What employer attitudes are desirable in confronting this situation? No employer, of course, can afford to pay people for doing no work. Yet especially within large corporations, it is often possible to readjust work assignments and workloads, to wait for people to retire, to change jobs voluntarily, to retrain employees for jobs within or outside the company, or even to introduce a 32-hour work week for all before laying off employees. Such benign employment practices have often been embraced from prudential motives to prevent a public and employee backlash against the introduction of computer technologies that eliminate jobs,[7] but moral considerations of human costs should be weighed even more heavily.

Customer Relations. There are questions about the public accountability of businesses using computer-based services. It can be either very difficult or relatively simple for a consumer to notice and correct computer errors or computer printed errors. For example, a grocery-store receipt can itemize items either by obscure symbols or by simple words understandable to a customer. The prices on cash register receipts have been shown to sometimes vary from those posted on the shelves. Here again moral reasons reinforce long-term good business sense in favoring policies that are beneficial to consumer needs and interests, making consumers feel less vulnerable.

Biased Software. In addition to computer hardware there is software, and programs can quite easily be biased, as can any form of communication or way of doing things. For example, a computerized study of the feasibility of constructing a nuclear power plant can easily become biased in one direction or another if the computer program is developed by a group entirely for or against nuclear energy.[8]

Stock Trading. Programmed trading is the automatic, hands-off, computer trading of stocks, futures, and options on the stock market. Did this practice contribute to the "meltdown on Black Monday" (October 19, 1987), when the U.S. stock market took a precipitous plunge, and should it be controlled? What assurances are there that NASDAQ, an electronic trading system linking

---

[7] Rob Kling, *Social Issues and Impacts of Computing* (Irvine, CA: University of California Press, 1979), 10.

[8] Deborah G. Johnson, *Computer Ethics,* 3rd ed. (Upper Saddle River, NJ: Prentice Hall, 2001).

510 stock traders, can prevent its members from exercising their power to manipulate the market when, as alleged, some have postponed requested purchases until after they have bought some shares of the same stock on their own, thereby raising the value of their newly acquired shares as well as increasing their commissions on the subsequent purchase of the shares requested by the customer? Controls have since been put into effect, but critics argue that more needs to be done.

Military Weapons. Military officials have often supported autonomous weapons that can be aimed and fired by onboard computers that make all necessary decisions, including enemy identification. The "launch-on-warning" policy for strategic missiles advanced by the U.S. military during the late 1980s could be considered an autonomous weapon. There is a dangerous instability in such automated defense systems, even if they are working perfectly. Even if all the nuclear warning software works without error, and the hardware is fail-safe, the combination of two such correctly functioning but opposing systems is unstable. This is because secrecy prevents either system from knowing exactly what the other is doing, which means that any input that could be interpreted as a danger signal must be responded to by an increase in readiness on the receiving side. That readiness, in turn, is monitored by the opposing side, which then steps up its readiness, and so on. This feedback loop triggers an escalating spiral. Does the possibility of an entirely unprovoked attack triggered by the interaction of two perfectly operating computer-based systems enhance security?[9]

Property

The most troublesome issues about property and computers fall under two general headings. The first is the use of computers in embezzlement and other forms of stealing money or financial assets. It is the most widely publicized form of computer crime and also the most morally clear-cut. The second set of issues concerns the theft of software and information. Here the issues are more complex.

Embezzlement. Computers are only incidentally involved when extortion is attempted through a phone that is part of a computerized telephone system.[10] By contrast, computers are

---

[9] Boris V. Rauschenbakh, "Computer War," in *Breakthrough: Emerging New Thinking,* ed. Anatoly Gromyko and Martin Hellman (New York: Walker, 1988).

[10] Rob Kling, "Computer Abuse and Computer Crime as Organizational Activities," *Computer/Law Journal* 2 (Spring 1980): 408.

centrally involved when an extortionist disguises his voice by means of a computer as he talks into a phone. And computers are even more centrally involved when an unauthorized person uses a telephone computer system to obtain private phone numbers, or when someone maliciously alters or scrambles the programming of a telephone computer.

Two factors make computers especially troublesome: (1) their speed and geographic coverage, which allows large numbers of people to be victimized, and (2) the difficulty of tracing the underlying transactions to apprehend the thieves. This problem is compounded when the communication lines linking the computers involved cross national boundaries.

Some of the most commonly discussed cases of computer abuse are instances of outright theft and fraud, of which there are many forms: (1) stealing or cheating by employees at work; (2) stealing by nonemployees or former employees; (3) stealing from or cheating clients and consumers; (4) violating contracts for computer sales or service; (5) conspiring to use computer networks to engage in widespread fraud. Especially alarming, the Internet has led to an explosion of identity theft, in which personal information is obtained and used to forge documents and commit fraud.

Public interest has often been drawn to the glamorous capers of computer criminals.[11] Enormous sums of money have been involved. The amount for an average computer-related embezzlement is 20 times the amount stolen in conventional embezzlement; many millions are often involved. Yet the giant thefts uncovered are believed to be only a small fraction of computer theft.

Crime by computer has proved to be unusually inviting. Computer crooks tend to be intelligent and to view their exploits as intellectual challenges. In addition, the computer terminal is both physically and psychologically far removed from face-to-face contact with the victims of the crimes perpetrated. Unlike violent criminals, computer criminals find it easy to deceive themselves into thinking they are not really hurting anyone, especially if they see their actions as nothing more than pranks. In addition, there are often inadequate safeguards against computer crime. The technology for preventing crime and catching criminals has

---

[11] Thomas Whiteside, *Computer Capers* (New York: Crowell, 1978); Tom Logsdon, *Computers and Social Controversy* (Potomac, MD: Computer Science Press, 1980), 163–64.

lagged behind the implementation of new computer applications. Computers reduce paperwork, but this has the drawback of removing the normal trail of written evidence involved in conventional white-collar crime (forgeries, receipts, etc.). Finally, the penalties for computer crime, as for white-collar crime in general, are mild compared with those for more conventional crimes.

Computer crime raises obvious moral concerns of honesty, integrity, and trust. It also forces a rethinking of public attitudes about crime and its punishment. Is it fair that the penalty for breaking into a gas station and stealing $100 should be the same as for embezzling $100,000 from a bank account? How should society weigh crimes of minor violence against nonviolent crimes involving huge sums of money?

The potential for computer crime should enter significantly into the thinking of engineers who design computers. In fact, protection against criminal abuse has become a major constraint for the effective and successful design of many computer systems and programs. Engineers must envisage not only the intended context in which the computer will be used, but both likely and possible abuses.

For some time, secret computer passwords have been used as a security feature. More recently introduced, and still of limited effectiveness, is data encryption. This technique is widely employed to prevent theft from funds transfer systems. In data encryption, messages are scrambled before transmission over communication lines and unscrambled after reception according to secret codes. Such devices, of course, require special precautions in maintaining confidentiality and security, and engineers have a major role to play in making recommendations in these areas. Of particular concern is the insistence by investigative agencies of the government that they be given access to decryption keys and that only prescribed codes be used. All of this tends to reduce the privacy of every user of the transmission system.

Data and Software. *Data,* in this context, refers to information stored in a computer, whether the information expresses facts or falsehoods. *Software* refers to programs that direct an electronic machine (hardware) to perform certain tasks, typically tasks involving problem solving. Programs have several aspects: (1) an algorithm, which explicitly states the steps in solving a problem; (2) a source code, which expresses the algorithm in a general computer language (such as Pascal, C, or FORTRAN); and (3) an object code, which translates a source code into the specific machine language of ones and zeros. Which of these aspects of computers are property, which can be privately owned, and which can be protected?

The question turns out to be surprisingly complex, and it forces us to clarify what property and property rights are (as we noted in Chapter 3). According to one primitive idea, persons' property is anything they create through their labor. John Locke, in the seventeenth century, developed this idea. According to Locke, we own our body and anything we "mix" with our body through labor. Locke had in mind a "state of nature" in which a person came to own a tree by either growing it or cutting it down, assuming no one else had previously done so. But once within a "state of society," property cannot be defined entirely by this simple idea. Property becomes primarily what laws define as the permissible use of things.

Laws define what can be owned, how exchanges of ownership may occur, and especially what ownership means in terms of the use of things of a given type. A car owner cannot drive on public roads until he or she satisfies laws about vehicle registration, insurance, and driver's licenses.[12] Again, a purchased book is the owner's, but that does not mean he or she can copy the entire book and sell it. What about ideas that a person or company develops for creating computers? Who owns them?

In the United States, computer hardware is protected by patent laws. Software can be protected by trade secret laws or by copyrights. Trade secret laws permit employers to require their employees not to divulge proprietary information. Obviously, trade secrets are useless once software is made publicly available as a marketed product. Here copyright laws offer the best protection.

Because of the newness of software, traditional laws are being extended to software gradually, often on a case-by-case basis. Generally, algorithms cannot be copyrighted. They are regarded as mathematical formulas that can be discovered but not owned. Laws stipulate that copyrighted material must be "intelligible," and the courts have tended to rule that object codes (written in machine language of ones and zeros) are not intelligible to humans and hence cannot be copyrighted. Source codes, however, are regarded as intelligible and can be copyrighted.

Patents on software are restricted to detailed coding sequences and other processes rather than final products. Not only are software patents difficult to obtain, they also create international disagreements because of differences in patent laws.

What does this mean? Does a company steal the property of a software producer if it buys one copy and then reproduces dozens of copies for its other employees? Yes, unless a special agreement has

---

[12] Deborah G. Johnson, *Computer Ethics.*

been reached with the software producer. Is making a dozen copies of a program borrowed from a friend for resale stealing? Yes.

Of course, one can always argue that particular laws are unjust, or, alternatively, that there are other overriding moral reasons that justify breaking a particular law. Nevertheless, the widespread practice of copying clearly denies the creators and producers of the programs the money to which they are entitled, and as such it is a form of theft. Forming user groups where self-generated programs are freely exchanged is another matter and a practice that should be encouraged.

Privacy

Storage, retrieval, and transmission of information using computers as data processors has revolutionized communication. Yet this very benefit poses moral threats to the right to privacy.[13] By making more data available to more people with more ease, computers make privacy more difficult to protect. Here we will discuss privacy and confidentiality for individuals, but the issues are similar for corporations.

Inappropriate Access. Imagine that you are arrested for a serious crime you did not commit—for example, murder or grand theft. Records of the arrest, any subsequent criminal charges, and information about you gathered for the trial proceedings might be placed on computer tapes easily accessible to any law enforcement officer in the country. Prospective employers doing security checks could gain access to the information. The record clearly indicates that you were found innocent legally. Nevertheless, that computerized record could constitute a standing bias against you for the rest of your life, at least in the eyes of many people with access to it.

The same bias could exist if you had actually committed some much less serious crime, say a misdemeanor. If you were arrested when you were 15 years old for drinking alcohol or swearing at an officer, for example, the record could stay with you. Or imagine that medical data about your visits to a psychiatrist during a period of depression could be accessed through a data bank. Or that erroneous data about a loan default were placed in a national credit data bank to which you had limited access. Or merely suppose that your tastes in magazine subscriptions were known easily to any employer or ad agency in the country and in the world.

---

[13] M. David Ermann, Mary B. Williams, and Michele S. Shauf, *Computers, Ethics, and Society,* 2nd ed. (New York: Oxford University Press, 1997).

Hackers. Finally there are "hackers," by which we mean that minority of computer enthusiasts sometimes called "crackers," who compulsively challenge any computer security system.[14] Some carry their art to the point of implanting "time bombs" or "Trojan horses" (unwanted codes that copy themselves into larger programs) that will "choke networks with dead-end tasks, spew out false information, erase files, and even destroy equipment."[15] This form of vandalism can be extremely harmful and is a straightforward violation of property rights, if only by reducing productivity by shutting down computer systems.

But suppose that the hacker's activities are limited to breaking into systems for shock value and a display of cunning. Is that so bad? After all, isn't it the responsibility of people to take appropriate steps to maintain their privacy, and isn't the hacker actually providing a stimulus for organizations to be more careful in protecting sensitive information? That is like arguing that it is all right to videotape the private activities of a neighbor who accidentally or carelessly leaves a window open. It is like arguing that if I do not invest in maximum security for my car that I authorize others to enter it.

Hackers sometimes employ a more extreme rationale in defending their activities. They contend that all information ought to be freely available, that no one should be allowed to own information, especially in a democratic society that respects individual rights to pursue knowledge. Essentially, this argument makes freedom of information paramount. Yet, there are at least three other important values that place legitimate limits on access to information: individual privacy, national security, and freedom within a capitalist economy to protect proprietary information essential in pursuing corporate goals.

Legal Responses. The potential abuses of information about us are unlimited and become more likely with the proliferation of access to that information. For this reason, a series of laws has been enacted.[16] For example, the 1970 Fair Credit Reporting Act restricted access to credit files. Information can be obtained only by consumer consent or a court order, or for a limited

---

[14] For a wider and positive meaning of "hacker," see Steven Levy, *Hackers* (New York: Penguin, 2001); and Pekka Himanen, *The Hacker Ethic* (New York: Random House, 2001).

[15] Eliot Marshall, "The Scourge of Computer Viruses," *Science* 240 (April 8, 1988): 133–34.

[16] James Rule, Douglas McAdam, Linda Stearns, and David Uglow, *The Politics of Privacy* (New York: New American Library, 1980).

range of valid credit checks needed in business, employment, and insurance transactions or investigations. The act also gave consumers the right to examine and challenge information about themselves contained in computerized files.

The Privacy Act of 1974 extended this right of inspection and error correction to federal government files. It also prohibited the information contained in government files from being used for purposes beyond those for which it was originally gathered unless such use was explicitly agreed to by the person whose file it is. Unfortunately, there is a loophole in the act that allows sharing of information among government agencies. Accordingly, there are now more than 100 separate computer matching programs to routinely pass data between agencies, greatly compromising personal privacy. However, many other laws have been passed and are being considered to extend the protection of individual privacy within private business and industry.

Such laws are expensive to implement, sometimes costing tens and hundreds of millions of dollars to enforce. They also lessen economic efficiency. In special circumstances they can have harmful effects on the public. There is little question, for example, that it would save lives if medical researchers had much freer access to confidential medical records. And it would be much more convenient to have one centralized National Data Center. This idea was proposed in the mid-1960s and is still alive in the minds of many. But privacy within a computerized world can apparently be protected only by making it inconvenient and expensive for others to gather information about us in data banks.

Additional Issues

Many of the issues in engineering ethics arise within the context of computer work. New variations or new difficulties may be involved, often owing to the high degree of job complexity and required technical proficiency introduced by computers. We provide some representative examples in the paragraphs that follow.

Computer Failures. Failures can occur because of errors in hardware or software. Hardware errors do not occur often, and when they do they usually do so quite obviously. An exception was Intel's highly touted Pentium chip, introduced in 1993. It produced very slight and rare errors in floating-point arithmetic. Perhaps more serious was the loss of confidence Intel suffered by not revealing the error before it was detected by a user.

Software errors are a different matter. They can be very serious indeed, as exemplified by the collapse of the computer-designed "space-frame" roof for the Hartford Civic Center in

1978, the deaths of several patients who received uncontrolled amounts of radiation in a radiation therapy machine between June 1985 and January 1987, and a major disruption of AT&T's computer-controlled long distance telephone system in 1990.

Errors can occur because of faulty logic in the design formulation, or they can be introduced in coding the machine instructions. Trial runs are absolutely essential to check out new programs, but if seasoned designers have already been replaced by canned programs and inexperienced engineers, the chance of noticing errors is slim. Perhaps this challenge to responsible management can be best met by engaging consultants to oversee the programming effort and to check the results of trial runs when the company does not have such in-house talent.

Computer Implementation. It should not be necessary to say so, but a changeover to a new computer system should never be attempted without having the old system still operational. Computer vendors who are too sure of their machines to recommend some redundancy during a changeover display enough hubris for it to qualify as one of the seven deadly sins. What can happen? Take the case of a bakery that had to file for bankruptcy after 75 successful years in the business. Part of the blame goes to slow summer sales, but there were extraordinary losses during the switch to a new computer system.

Health Conditions. Engineers who supervise computer personnel or design computer terminals should check that ergonomic considerations are in effect to reduce back problems, provide wrist support, offer good keyboard layouts to prevent carpal tunnel syndrome, and offer good lighting and flicker control.

Discussion Questions

1. Consider an engineer who develops a program used as a tool in developing other programs assigned to her. Subsequently she changes jobs and takes the only copy of the first program with her for use on her new job. Suppose first that the program was developed on company time under the first employer's explicit directives. Taking it to a new job without the original employer's consent would be a violation of that employer's right to the product (and possibly a breach of confidentiality). As a variant situation, however, suppose the program was not written under direct assignment from the first employer, but was undertaken by the engineer at her own discretion to help her on her regular work assignments. Suppose also that to a large extent the program was developed on her own time on weekends, although she

did use the employer's facilities and computer services. Did the employer own or partially own the program? Was she required to obtain the employer's permission before using it on the new job?[17]

2. Dependence on computers has intensified the division of labor within engineering. For example, civil engineers designing a flood control system have to rely on information and programs obtained from systems analysts and implemented by computer programmers.

   Suppose the systems analysts refuse to assume any moral or legal responsibility for the safety of the people affected by the flood control plans, arguing that they are merely providing tools whose use is entirely up to the engineers. Should the civil engineers be held accountable for any harm caused by poor computer programs? Presumably their accountability does extend to errors resulting from their own inadequate specifications that they supply to the computer experts. Yet should not the engineers also be expected to contract with computer specialists who agree to be partially accountable for the end-use effects of their programs?[18]

3. An engineer working as a computer programmer played a minor role in developing a computer system for a state department of health. The system stored medical information on individuals identified by name. Through no fault of the engineer, few controls had been placed on the system to limit easy access to it by unauthorized people. Upon learning of this, the engineer first informed his supervisor and then higher management, all of whom refused to do anything about the situation because of the anticipated expense required to correct it. In violation of the rules for using the system, the programmer very easily obtained a copy of his own medical records. He then sent them to a state legislator as evidence for his claims that the right of citizens to confidentiality regarding such information was threatened by the system. Was his behavior improper? Was his subsequent firing justified?[19]

4. A project leader working for a large retail business was assigned the task of developing a customer billing and credit system. The budget assigned for the project appeared at first to be adequate. Yet by the time the system was half completed it was clear the funds were not nearly enough. The project leader asked for more money, but the request was denied. He fully informed manage-

---

[17] Donn B. Parker, *Ethical Conflicts in Computer Science and Technology* (Arlington, Virginia: AFIPS Press, 1979), 72–74. Case studies adapted in the text with permission of author and publisher.

[18] Ibid., 34–38.

[19] Ibid., 90–93.

ment of the serious problems that were likely to occur if he had to stay within the original budget. He would be forced to omit several important program functions for convenience and safety: for example, efficient detection and correction mechanisms for errors, automatic handling and reporting of special customer exceptions, and audit controls. Management insisted that these functions could be added after the more minimal system was produced and installed in stores. Working under direct orders, the project leader completed the minimal system, only to find his worst fears realized after it was installed. Numerous customers were given incorrect billings or ones they could not understand. It was easy for retail salespersons to take advantage of the system to steal from the company, and several did so. Within a year the company's profits and business were beginning to drop. This led to middle-level management changes, and the project leader found himself blamed for designing an inadequate system.

Did the project leader have an obligation either to clients or to the company to act differently than he did? Did he have a moral right to take further steps in support of his original request or later to protect himself from managerial sanctions?[20]

5. A team of engineers and biomedical computer scientists develop a system for identifying people from a distance of up to 200 meters. A short tube attached to a sophisticated receiver and computer, and aimed at a person's head, reads the individual's unique pattern of brain waves when standard words are spoken. The team patents the invention and forms a company to manufacture and sell it. The device is an immediate success within the banking industry. It is used to secretly verify the identification of customers at tellers' windows. The scientists and engineers, however, disavow any responsibility for such uses of the device without customer notification or consent. They contend that the companies that buy the product are responsible for its use. They also refuse to be involved in notifying public representatives about the product's availability and the way it is being used.

Does employing the device without customer awareness violate the right to privacy or to informed consent? Do the engineers and scientists perhaps have a moral obligation to market the product with suggested guidelines for its ethical use? Should they be involved in public discussions about permissible ways of using it?[21] (Retina scan identification systems using laser beams are already in use. An example would be to determine if a person using a particular computer is authorized to use it.)

---

[20] Ibid., 109–11.
[21] Ibid., 126–28.

6. The following warning to parents whose children use home computers was carried by the Associated Press: "In recent years more sexually oriented materials have been showing up for home computers—some on floppy disks with X-rated artwork and games, and other accessed by phone lines from electronic bulletin boards . . . with names like Cucumber, . . . Orgy, Nude pics, Porno, Xpics, and Slave."[22]

Discuss the ethical issues raised by pornography in this new medium, as well as the issues raised by racist, sexist, and libelous (false and defamatory) statements. How can access be denied to children? Should there be controls for adults? Already there are thousands of bulletin boards, largely because it is so easy and inexpensive to create them. Should bulletin board operators be held liable for failing to filter illegal forms of verbal assaults, even if that forces them to buy liability insurance and thereby raise the costs of creating bulletin boards?

7. Write a short research paper exploring the threats to privacy posed by data banks. In your essay, comment on some specific advantages and disadvantages of having one centralized national data bank that pools all available government information on citizens.

---

[22] Associated Press, *Los Angeles Times,* December 25, 1987, part 1, 47.

CHAPTER 9

Environmental Ethics

Human life is possible because of the greenhouse effect, in which atmospheric gases such as water vapor and carbon dioxide block solar energy from escaping, after being reflected from the earth's surface. In 1988, however, NASA scientist James Hansen warned that the greenhouse effect is accelerating owing to human burning of fossil fuels that increase levels of greenhouse gases such as carbon dioxide (CO_2). The change is small, but even a few degrees of global warming could melt enough of the polar ice caps to raise the oceans enough to cause severe flood damage. Other effects include major disruptions in weather patterns, such as increased drought, major shifts in rain patterns, and increased severity of hurricanes caused by rising ocean temperatures.

For years, the complexity of the issue divided scientific opinion, but an emerging consensus led to the 1997 Kyoto agreement, signed by 150 governments, to reduce carbon emissions to 5.2 percent below 1990 levels by 2012. The United States abandoned the agreement on the grounds that it was antithetical to American business and unfair in not placing stronger requirements on developing countries (even though the United States was by far the major source of carbon emissions per capita). By 2007, however, the evidence was overwhelming that global warming had become an urgent problem. By then, as well, public awareness about the problem had been raised by environmental activists, by Al Gore's Academy-Award winning film *An Inconvenient Truth*, and perhaps most of all by Hurricane Katrina that tragically demonstrated how rising ocean temperatures can add to the fury of hurricanes. Simultaneously, the Iraq War intensified concerns about overdependence on foreign oil and the general desirability of increasing the use of renewable energy sources. Corporations that had resisted recognition of the problem and in some cases funded groups that spread disinformation about it, abruptly reversed direction, some out of genuine moral concern and some to position themselves to have greater input into forging new government regulations.

In addition to global warming, environmental challenges confront us at every turn, including myriad forms of pollution, human-population growth, extinction of species, destruction of ecosystems, depletion of natural resources, and nuclear waste. Today there is a wide consensus that we need concerted environmental responses that combine economic realism with ecological awareness. For their part, many engineers are now showing leadership in advancing ecological awareness. In this chapter, we discuss some ways in which this responsibility for the environment is shared by engineers, industry, government, and the public. We also introduce some perspectives developed in the new field of environmental ethics that enter into engineers' personal commitments and ideals.

9.1 Engineering, Ecology, and Economics

Like the word *ethics,* the expression *environmental ethics* can have several meanings. We use the expression to refer to (1) the study of moral issues concerning the environment, and (2) moral perspectives on those issues.

The Invisible Hand and the Commons

Two powerful metaphors have dominated thinking about the environment: the invisible hand and the tragedy of the commons. Both metaphors are used to highlight unintentional impacts of the marketplace on the environment, but one is optimistic and the other is cautionary about those impacts. Each contains a large part of the truth, and they need to be reconciled and balanced.

The first metaphor was set forth by Adam Smith in 1776 in *The Wealth of Nations,* the founding text of modern economics. Smith conceived of an invisible (and divine) hand governing the marketplace in a seemingly paradoxical manner. According to Smith, businesspersons think only of their own self-interest: "It is not from the benevolence of the butcher, the brewer, or the baker, that we expect our dinner, but from their regard to their own interest."[1] Yet, although "he intends only his own gain," he is "led by an invisible hand to promote an end which was no part of his intention. . . . By pursuing his own interest he frequently promotes that of the society more effectually than when he really intends to promote it. I have never known much good done by those who affected to trade for the public good."[2]

---

[1] Adam Smith, *An Inquiry into the Nature and Causes of the Wealth of Nations,* vol. 1 (New York: Oxford University Press, 1976), 26–27.
[2] Ibid., 456.

In fact, professionals and many businesspersons do profess to "trade for the public good," claiming a commitment to hold paramount the safety, health, and welfare of the public. Although they are predominantly motivated by self-interest, they also have genuine moral concern for others.[3] Nevertheless, Smith's metaphor of the invisible hand contains a large element of truth. By pursuing self-interest, the businessperson, as entrepreneur, creates new companies that provide goods and services for consumers. Moreover, competition pressures corporations to continually improve the quality of their products and to lower prices, again benefiting consumers. In addition, new jobs are created for employees and suppliers, and the wealth generated benefits the wider community through consumerism, taxes, and philanthropy.

Despite its large element of truth, the invisible hand metaphor does not adequately take account of damage to the environment. Writing in the eighteenth century, with its seemingly infinite natural resources, Adam Smith could not have foreseen the cumulative impact of expanding populations, unregulated capitalism, and market "externalities"—that is, economic impacts not included in the cost of products. Regarding the environment, most of these are negative externalities—pollution, destruction of natural habitats, depletion of shared resources, and other unintended and often unappreciated damage to "common" resources.

This damage is the topic of the second metaphor, which is rooted in Aristotle's observation that we tend to be thoughtless about things we do not own individually and which seem to be in unlimited supply. William Foster Lloyd was also an astute observer of this phenomenon. In 1833 he described what the ecologist Garrett Hardin would later call "the tragedy of the commons."[4] Lloyd observed that cattle in the common pasture of a village were more stunted than those kept on private land. The common fields were themselves more worn than private pastures. His explanation began with the premise that individual farmers are understandably motivated by self-interest to enlarge their common-pasture herd by one or two cows, especially given that each act taken by itself does negligible damage. Yet, when all the farmers behave this way, in the absence of laws constraining them, the result is the tragedy of overgrazing that harms everyone.

---

[3] In *The Theory of Moral Sentiments,* Adam Smith also stressed human capacities for altruism. Scholars struggle with how to reconcile the seeming contradiction in Smith's outlook. See Patricia H. Werhane, *Adam Smith and His Legacy for Modern Capitalism* (New York: Oxford University Press, 1991).

[4] Garrett Hardin, *Exploring New Ethics for Survival* (New York: Viking, 1968), 254.

The same kind of competitive, unmalicious but unthinking, exploitation arises with all natural resources held in common: air, land, forests, lakes, oceans, endangered species, and indeed the entire biosphere. Hence, the tragedy of the commons remains a powerful image in thinking about environmental challenges in today's era of increasing population and decreasing natural resources. Its very simplicity, however, belies the complexity of many issues concerning ecosystems and the biosphere. Ecosystems are systems of living organisms interacting with their environment—for example, within deserts, oceans, rivers, and forests. The biosphere is the entirety of the land, water, and atmosphere in which organisms live. Ecosystems and the biosphere are themselves interconnected and do not respect national boundaries. There is need for multifaceted and often concerted environmental responses by engineers, corporations, government, market mechanisms, local communities, and social activists.

Engineers: Sustainable Development

Ali Ansari, a scholar in India, suggests that there is a "standard engineering worldview—that of a mechanical universe," which is at odds with mainstream "organic" environmental thought.[5] According to Ansari, central to the engineering view is "techno-think," which "implicitly assumes that things can be understood by analyzing them and, if something goes wrong, can be fixed." In contrast, "green philosophy" "demands humility, respect, and sensitivity towards the natural world."

We believe there is a tension, not a dichotomy, between techno-think and green philosophy, as Ansari defines them. It is true that historically engineers were not as responsible concerning the environment as they should have been, but in that respect they simply reflected attitudes predominant in society. The U.S. environmental movement that emerged from the 1960s began a social transformation that has influenced engineers as much as other populations, and more than most professions. Furthermore, there is no single canonical professional attitude or philosophical "green" attitude. Individual engineers, like individuals in all professions, differ considerably in their views, including their broader holistic views about the environment. What is important is that all engineers should reflect seriously on environmental values and how they can best integrate them into understanding and solving problems. In doing so, as Sarah Kuhn points out in replying to Ansari, engineers should also be able to "work in an organizational context in which an eco-friendly approach is

---

[5] Ali Ansari, "The Greening of Engineers: A Cross-Cultural Experience," *Science and Engineering Ethics* 7, no. 1 (2002): 105, 115.

valued and supported with the tools, information, and incentives necessary to succeed. Beyond that, they must work in a market that rewards sustainable products and processes, and in a policy context that encourages, or at least does not discourage, environmental protection."[6]

In many respects, engineers are singularly well-placed to make environmental contributions. They can encourage and nudge corporations in the direction of greater environmental concern, finding ways to make that concern economically feasible. At the very least, they can help ensure that corporations obey applicable laws. In all these endeavors, they benefit from a supportive code of ethics stating the shared responsibilities of the profession.

Increasingly, engineering codes of ethics explicitly refer to environmental responsibilities under the heading of "sustainable development."[7] In the United States, a first important step occurred in 1977 when the American Society of Civil Engineers (ASCE) introduced into its code the statement "Engineers should be committed to improving the environment to enhance the quality of life." "Should" indicates the desirability of doing so, although (in contrast to "shall") it does not indicate something mandatory or enforceable. Still, the mere mention of the environment was a breakthrough. Two decades later, in 1997, ASCE's fundamental canon has changed from recommendations ("should") to requirements ("shall"): "Engineers shall hold paramount the safety, health and welfare of the public and shall strive to comply with the principles of sustainable development in the performance of their professional duties." Additional requirements are added that require notifying "proper authorities" when the principles of sustainable development are violated by employers, clients, and other firms.

What is "sustainable development" (sometimes shortened to "sustainability")? The term was introduced in the 1970s, but it became popular during the 1980s and 1990s, especially since the publication in 1987 of *Our Common Future,* produced by the United Nations in its World Commission on Environment and Development (also called the Brundtland Report).[8] Put negatively, the term was invented to underscore how current patterns of economic activity and growth cannot be sustained as populations grow, technologies are extended to developing countries,

---

[6] Sarah Kuhn, "Commentary On: The Greening of Engineers: A Cross-Cultural Experience," *Science and Engineering Ethics* 7, no. 1 (2001): 124.

[7] The following discussion is based on P. Aarne Vesilind and Alastair S. Gunn, *Engineering, Ethics, and the Environment* (New York: Cambridge University Press, 1998), 48–65.

[8] Alan Holland, "Sustainability," in *A Companion to Environmental Philosophy,* ed. Dale Jamieson (Malden, MA: Blackwell Publishing, 2003), 390–401.

and the environment is increasingly harmed. Put positively, the term implies the crucial need for new economic patterns and products that are sustainable, that is, compatible with both ongoing technological development and protection of the environment. As such, the term suggests a compromise stance between advocates of traditional economic development that neglected the environment, and critics who warned of an environmental crisis: Economic development is essential, but it must be sustainable into the future. The compromise is somewhat uneasy, however, for different groups understand its meaning in different ways.

In *Our Common Future,* sustainable development is defined as "development that meets the needs of the present without compromising the ability of future generations to meet their own needs."[9] This statement emphasizes *inter*generational justice—balancing the needs of living populations against those of future generations. The document also calls for greater *intra*-generational justice—justice in overcoming poverty among living populations, for conserving natural resources, and for keeping populations at sustainable levels. In tune with these themes, ASCE defines *sustainable development* as "a process of change in which the direction of investment, the orientation of technology, the allocation of resources, and the development and functioning of institutions [is directed] to meet present needs and aspirations without endangering the capacity of natural systems to absorb the effects of human activities, and without compromising the ability of future generations to meet their own needs and aspirations."[10]

Critics charge that the term "sustainable development" is a ruse that conceals business as usual under the guise of environmental commitment.[11] Undoubtedly, sometimes it is, but there are also many individuals and corporations who use the term to convey genuine environmental concern. It seems likely the term will continue to be a rallying point for attempts to find common ground in thinking about how to integrate economic and ecological concerns.

Corporations: Environmental Leadership

In the present climate, it is good business for a corporation to be perceived by the public as environmentally responsible, indeed as

---

[9] World Commission on Environment and Development, *Our Common Future* (Oxford: Oxford University Press, 1987).

[10] American Society of Civil Engineers, "The Role of the Engineer in Sustainable Development," www.asce.org.

[11] Aidan Davison, *Technology and the Contested Meanings of Sustainability* (Albany, NY: State University of New York Press, 2001).

a leader in finding creative solutions. This is true of corporations of all sizes.

General Electric (GE) is a good example of a large corporation that took initiative in recent years, convinced that in today's business climate ecology is good economics.[12] Under a program called Ecomagination, GE drew together, expanded, and built on its businesses that were environmentally friendly, compared to current competitors. New investments were made, such as purchasing for $358 million the wind-turbine business of Enron (during its bankruptcy). It also intensified its research and development in biofuels and other renewable energy resources. In addition, it was one of the corporations that showed leadership in 2007 by urging the federal government to take action on climate change.

As an example of a moderate-size company, we cite the Solar Electric Light Fund (SELF), which provides solar energy in developing countries, beginning with rural areas in South Africa and China.[13] Neville Williams had several core convictions when he founded SELF. He was convinced that replicating the traditional fossil fuel and grid-distribution system of energy distribution would do enormous environmental damage. He also knew that marketing environmentally friendly sources of power was crucial now, not only because of enormous demand but also because the first system in place could shape the future trend. And he knew that although financing was a huge obstacle for families, it was nevertheless important for them to pay for the energy and accept responsibility for taking care of the technology once it was implemented. A simple solar technology, combined with reasonable financing plans, proved realistic. Photovoltaic units for homes that provided 20 years of energy could be marketed at $500. An innovative funding system was devised whereby grant money provided loans for initial sales, and then payments on the loans were used to finance additional loans.

Government: Technology Assessment, Incentives, Taxes

Government laws and regulations are understandably the lightning rod in environmental controversies. Few would question the need for the force of law in setting firm guidelines regarding the degradation of the "commons," especially in limiting the excesses of self-seeking while establishing fair "rules to play by." Yet, how much law, of what sort, and to what ends are matters of continual

---

[12] "Cleaning Up: A Special Report on Business and Climate Change," *Economist* (June 2, 2007), 12–14.

[13] Michael E. Gorman, et al., "Toward a Sustainable Tomorrow," in *Business Consumption: Environmental Ethics and the Global Economy,* ed. Laura Westra and Patricia H. Werhane (Lanham, MD: Rowman & Littlefield, 1998), 333–38.

disagreement. In the United States, landmark environmental legislation at the national level in the United States began in 1969 with passage of the National Environmental Policy Act, which requires environmental impact statements for federally funded projects affecting the environment. Other key legislation quickly followed, including the Occupational Safety and Health Act (1970), the Clean Air Act (1970), the Clean Water Act (1972), and the Toxic Substances Control Act (1976). These and subsequent legislative acts involved heated controversy at several stages: in passing laws, in developing and managing enforcement procedures, and in modifying laws to take account of unforeseen problems.

Government and international agreements among governments are essential in tackling global warming, especially in putting a price on CO_2 emissions. This can be accomplished in several ways. One option is to establish standards and requirements for energy efficiency in vehicles and new buildings, and to ban incandescent light bulbs. Another option is to tax CO_2 emissions, a traditional approach that is relatively straightforward and sends a clear message about penalties for pollution. A third option is establish a "cap-and-trade system," of the sort adopted in Europe following the Kyoto Treaty. This option allows carbon emission, or rather documented noncarbon emission, to be bought and sold as a commodity. The drawback has been the strong fluctuations in the market that create uncertainties for business, but the approach is perceived by many as a preferable "free-market" solution, which brings us to the next approach.

Market Mechanisms: Internalizing Costs

Democratic controls take many forms beyond passing laws. One such option is internalizing costs of harm to the environment. When we are told how efficient and cheap many of our products and processes are—from agriculture to the manufacture of plastics—the figures usually include only the direct costs of labor, raw materials, and the use of facilities. If we are quoted a dollar figure, it is at best an approximation of the price. The true cost would have to include many indirect factors such as the effects of pollution, the depletion of energy and raw materials, disposal, and social costs. If these, or an approximation of them, were internalized (added to the price) then those for whose benefit the environmental degradation had occurred could be charged directly for corrective actions.

Taxpayers are revolting against higher levies, so the method of having the user of a particular service or product pay for all its costs is gaining more favor. The engineer must join with the economist, the scientist, the lawyer, and the politician in an effort to find acceptable mechanisms for pricing and releasing products so that the environment is protected through truly self-correcting procedures rather than adequate-appearing yet often circumventable laws. A working example is the tax imposed by governments in Europe on products and packaging that impose a burden on public garbage disposal or recycling facilities. The manufacturer prepays the tax and certifies so on the product or wrapper.

Communities: Preventing Natural Disasters

Communities at the local and even state level have special responsibility to conserve natural resources and beauty for future generations. They have special responsibility, as well, for preventing natural events—such as hurricanes, floods, fires, and earthquakes—from becoming disasters. There are four sets of measures communities can take to avert or mitigate disasters.

One set of defensive measures consists of restrictions or requirements imposed on human habitat. For instance, homes should not be built in floodplains, homes in prairie country should have tornado shelters, hillsides should be stabilized to prevent landslides, structures should be able to withstand earthquakes and heavy weather, roof coverings should be made from nonflammable materials, and roof overhangs should be fashioned so flying embers will not be trapped. These are not exorbitant regulations, but merely reminders to developers and builders to do what their profession expects them to do anyway.

A second set of measures consists of strengthening the lifelines for essential utilities such as water (especially for fire fighting) and electricity. A third category encompasses special-purpose defensive structures that would include dams, dikes, breakwaters, avalanche barriers, and means to keep floodwaters from damaging low-lying sewage plants placed where gravity will take a community's effluents. A fourth set of measures should assure safe exits in the form of roads and passages designed as escape routes, structures designated as emergency shelters, adequate clinical facilities, and agreements with neighboring communities for sharing resources in emergencies.

When disasters do occur, lessons can be learned, rather than shrugged aside by a disbelief that the event could occur again—"Lightning never strikes twice in the same place," and "Another 100-year flood is about that far away"—or by a belief that government would once more hand out disaster relief payments.

Communities show leadership, for example, when they develop programs that encourage recycling, often in conjunction with state governments. In May 2003, the California Department of Conservation launched an awareness campaign to encourage the recycling of plastic bottles. In one year, four billion plastic bottles were sold in California, but only a third of those were recycled.

The rest ended up in landfills where they are not biodegradable. Worse, they were often incinerated, which released toxic fumes. A human-made disaster is in the making: "If the problem continues, enough water bottles will be thrown in the state's trash dumps over the next five years to create a two-lane, 6-inch-deep highway of plastic along the entire California coast."[14] Cities can alleviate the problem by making recycle bins readily available. In addition, the legislature is seeking to raise the cash refund on bottles. At the national level, there is hope that consumer habits might be modified by extensive news reports that tap water is as healthy as most bottled water, indeed that much bottled water just is tap water.

Social Activists

Social activism by concerned citizens has played a key role in raising public awareness. As examples, we cite Rachel Carson, Sherwood Rowland, and Engineers Without Borders.

In the United States the environmental movement had many roots, but its catalyst was Rachel Carson's 1962 book *Silent Spring*. Carson made a compelling case that pesticides, in particular dichlorodiphenyltrichloroethane (DDT), were killing creatures beyond their intended target, insects. DDT is a broad-spectrum and highly toxic insecticide that kills a variety of insects. It also persists in the environment by being soluble in fat, and hence storable in animal tissue, but not soluble in water, so that it is not flushed out of organisms. As a result, DDT enters into the food chain at all levels, with increasing concentrations in animals at the higher end of the chain.

To the public, Carson had scientific credibility, having earned a bachelor's degree in biology and a master's degree in zoology and then spending a career working for the Fish and Wildlife Service. But many other scientists with stronger credentials had been warning of the dangers of DDT for nearly two decades. Carson was unique because her prose combined scientific precision, poetic expression, and a trenchant argument understandable by the general public. Critics, especially chemical companies, were less sympathetic. She was patronized, mocked, and reviled as a sinister force that threatened American industry.[15] If today she is an American icon, it is in large measure because of the courage of her convictions in confronting a hostile establishment. At the same time, we have since gained new knowledge that balances Carson's insights, in particular an appreciation that DDT remains a valuable way to fight malaria by killing the mosquitoes

---

[14] Miguel Bustillo, "Water Bottles Are Creating a Flood of Waste," *Los Angeles Times,* May 28, 2003, B-1 and B-7.

[15] Frank Graham, Jr., *Since Silent Spring* (Boston: Houghton Mifflin, 1970).

that spread it. DDT has been banned in the United States and other western nations since the 1970s, but when its use in Madagascar was suspended in 1986, 100,000 deaths from malaria occurred, leading to its immediate reuse.[16]

Our second example is Professor Sherwood Rowland at the University of California, Irvine, who also confronted the wrath of an entire industry following the publication of a 1974 essay in *Nature,* coauthored with Mario Molina, identifying the depletion of the ozone layer by chlorofluorocarbons (CFCs).[17] Rowland and Molina, building on the work of Paul Crutzen, argued that CFCs were rising 15 miles to the stratosphere and beyond, where they broke down ozone (O_3). The ozone layer protects the entire planet from deadly ultraviolet (UV) radiation; although it is a relatively thin and diffuse layer, it is critical for protecting nearly all life forms. CFC gases, such as freon, are synthetic chemicals that since the 1930s had been widely used in refrigerators and air conditioners, and also as propellants in aerosol spray cans. Hence, in setting forth a scientific argument, the authors set off a firestorm of protest from industry, and Rowland in particular spent much of the next decade countering criticism. NASA tests in 1987 confirmed what Rowland and Molina had argued by identifying huge areas of thinning of the ozone layer. In the same year, with unprecedented speed, the Montreal Protocol, signed by the main producers and users of CFCs, mandated the phaseout of CFCs by 2000. The danger persists, however, because the CFCs already produced will go on interacting with ozone for decades, requiring addition UV protection by sunbathers to prevent deadly skin cancers.

Rowland, Molina, and Paul Crutzen (a Dutch scientist who showed that nitric oxide [NO] and nitrogen dioxide [NO_2] react catalytically with ozone) were awarded the Nobel Prize in Chemistry in 1995, the first time the prize was given for applied environmental science.

Finally, social activism occurs at many levels, the micro as well as macro. In 2002, Bernard Amadei founded Engineers Without Borders, a philanthropic organization with a core commitment to help disadvantaged communities through environmentally sensitive and sustainable engineering projects.[18] At present, more than 250 projects in 43 nations are underway, for example in creating energy efficient water pumps and electrical systems that are sufficiently simple to be maintained by local residents when

[16] Greg Pearson and A. Thomas Young, eds., *Technically Speaking* (Washington, DC: National Academy Press, 2002), 19.

[17] Mary H. Cooper, "Ozone Depletion," *CQ Researcher* (April 3, 1992).

[18] Engineers Without Borders, www.ewb-international.org.

engineers move on to other projects, or return to their work at home.

Two Corps Cases

We conclude with two cases involving the U.S. Army Corps of Engineers, one a promising success and the other the worst natural disaster in U.S. history, that illustrate the need for developing reasonable compromises and consensus between local, state, and federal groups.

The promising success is work on a 55-mile waterway through the Napa Valley in California, which for decades had been a battleground between humans and nature. A series of earthen levees were used to redirect and constrain the natural flow of water, and low concrete bridges spanned the river, but these makeshift controls failed to prevent periodic floods. A 1986 flood alone caused $100 million in property damage, killed three people, and forced the evacuation of 5000 others. In July 2000, groundbreaking ceremonies took place on a project to restore the waterway into a "living river," with natural floodplains, wetlands, and other natural habitat.[19] The restoration project was innovative in the way it combined ecological and economic goals. Instead of imposing further constraints on the river, the project was to restore the river to something closer to its natural state. Furthermore, although initial costs would be much higher, even after the savings from conforming to regulations about preserving wetlands and endangered species, there were counterbalancing long-term economic benefits. They included increased tourism because of more scenic countryside, elevated property values, and lower home insurance costs as a result of lessened flood risks.

The solution was also innovative in the role played by the U.S. Army Corps of Engineers. Initially, the Corps proposed a deeper concrete channel with higher concrete walls and concrete steps. This proposal reflected traditional Corps thinking.[20] The Corps is comprised of talented and conscientious engineers, with some 30,000 civilian employees directed by approximately 200 Army officers who have enormous power, reporting directly to Congress through the Office of Management and Budget. Yet, the Corps had acquired a controversial tradition of reengineering nature with a preference for straight-line, concrete structures

---

[19] Gretchen C. Daily and Katherine Ellison, *The New Economy of Nature: The Quest to Make Conservation Profitable* (Washington, DC: Island Press, 2002), 87–108.

[20] P. Aarne Vesilind, "Decision Making in the Corps of Engineers: The B. Everett Jordan Lake and Dam," in *Engineering, Ethics, and the Environment*, ed. P. Aarne Vesilind and Alastair S. Gunn (Cambridge: Cambridge University Press, 1998), 171–77; Arthur E. Morgan, *Dams and Other Disasters* (Boston: Porter Sargent, 1971), 370–89.

that placed environmental concerns, not to mention the desires of local communities, a distant second.

The local community rejected the Corps' initial plan. Members of the community were asked to pay higher taxes to help fund a concrete eyesore slicing through miles of beautiful farmland and also the middle of Napa City. Activists rallied support from citizens and various state agencies for an alternative plan involving multiple compromises. Eventually the plan also gained the support of the Corps, which was essential because its federal money funded most of the project.

According to the compromise plan, the Corps will destroy current levees and nine bridges, rebuilding five bridges at higher levels after rerouting the water closer to its original state. The local community will accept a one-half cent increase in sales taxes, and major sacrifices will be required of some homes and businesses that require relocation. But the long-term economic benefits promise to outweigh the sacrifice. Of course, no one can foresee the exact costs and benefits, and critics argue that global climate changes might lead to greater flood problems than anyone can predict. The project is truly a social—and environmental—experiment. But there is a basis for hope in this "collaboration with nature," especially because of how it integrates enlightened self-interest, ecological awareness, and engineering expertise.

Contrast this case with the battle against nature won decisively by Hurricane Katrina in 2005 (Figure 9–1).[21] More than 1,100 people died in Louisiana and hundreds more in surrounding states. In addition, 1.7 million were evacuated, resulting in hundreds of thousands of evacuated people permanently dispersed throughout the United States. More than 100,000 people who had no car or access to transportation were trapped. Most of them were poor and people of color; 1,700 were in hospitals; 2,430 children were separated from their parents; and nearly all were left without food, water, electricity, or shelter in the days that followed. In addition, huge numbers of pets were killed, for authorities refused to allow them into the rescue vehicles, resulting in additional human deaths as some individuals chose to stay with their pets rather than leave them to die. (Rescue policies for pets have been revised in light of Katrina.)

The causes of the disaster were multiple. They included the perfect storm that hit with the fury of 155–miles-per-hour winds and with uncanny precision. They included the relentless incursion of homes and roads into the wetlands that provided a natural

---

[21] Douglas Brinkley, *The Great Deluge: Hurricane Katrina, New Orleans, and the Mississippi Gulf Coast* (New York: William Morrow, 2006).

Figure 9-1
August 30, 2005.
An arial view of a
flooded section of New
Orleans after the levee
broke

Jocelyn Augustino/FEMA

buffer against hurricanes. They included utter failure of the city and region to develop and implement disaster preparedness plans, especially for the 100,000 individuals who lacked cars or other means of escape. Equally egregious and shameful failures occurred at the state and federal level, as the officials at the Federal Emergency Management Agency (FEMA) and the Department of Homeland Security, of which FEMA had become a part, were at times utterly disconnected from the tragedy they were responsible for dealing with. And the causes included shoddy engineering.

New Orleans is built on a swamp and mostly below sea level. The Corps is primarily responsible for the complex 350-mile system of dikes and levees protecting New Orleans from the Mississippi River and Lake Pontchartrain, as well as from hurricanes. It only took only a few levees giving way to flood the city, in most areas with 10 to 15 feet of what would come to be known as a "toxic gumbo." Lake Pontchartrain simply emptied into the city until it stabilized at the level of city flooding. The failure of the levees was not caused primarily by water overtopping them. Instead, the rising and wind-churned water pushed the levees enough to open gaps between the foundations of the levees and the soft clay beneath them; then, water quickly eroded the clay. Too many levees were poorly designed and inadequately inspected and maintained.

At first, officials claimed that the disaster was not foreseeable. In fact, the dangers were well known for many years. Although the Corps initially denied the tragedy could have been prevented, they were forced to acknowledge that they had conducted studies twenty years earlier that revealed exactly this scenario.[22] And the lack of disaster preparedness, at all levels of government, were well known among officials prior to Katrina.

Discussion Questions

1. Identify and comment on the importance of each of the environmental impacts described in the following passage: "The Swedish company IKEA, the world's largest furniture and home furnishings retailer, has adopted a global corporate policy that prohibits the use of old-growth forest wood or tropical wood in its furniture. All timber must come from sustainably managed forests. IKEA has eliminated the use of chlorine in its catalog paper, uses 100 percent recycled paper fibers, and is committed to eliminating waste in its retail stores. The 'Trash is Cash' program has transformed the thinking of retail store workers to see trash as a revenue-generating resource."[23]

2. Most companies want to have a reputation for environmental responsibility, but there are different "shades of green" in their commitments.[24] They include (1) "light green"—compliance with the law; (2) "market green"—seeking competitive advantage by attending to customer preferences; (3) "stakeholder green"—responding to and fostering environmental concern in the stakeholders of the corporation, including suppliers, employees, and stockholders; and (4) "dark green"—creating products and using procedures that include respect for nature as having inherent worth. Which of these shades of green would you ascribe to GE and to SELF?

3. Identify and discuss the moral issues involved in the following case.

   The great marshes of southern Florida have attracted farmers and real estate developers since the beginning of the century. When drained, they present valuable ground. From 1909 to 1912 a fraudulent land development scheme was attempted in collusion with the U.S. Secretary of Agriculture. Arthur Morgan blew the whistle on that situation, jeopardizing not only his own position as a supervising drainage engineer with the U.S.

---

[22] John McQuaid and Mark Schleifstein, *Path of Destruction* (New York: Little, Brown and Company, 2006), 342–43.

[23] Andrea Larson, "Consuming Oneself: The Dynamics of Consumption," in *The Business of Consumption: Environmental Ethics and the Global Economy,* ed. Laura Westra and Patricia H. Werhane, 320–21.

[24] R. Edward Freeman, Jessica Pierce, and Richard Dodd, "Shades of Green: Business, Ethics, and the Environment," in *The Business of Consumption: Environmental Ethics and the Global Economy,* ed. Laura Westra and Patricia H. Werhane (Lanham, MD: Rowman & Littlefield Publishers, 1998), 339–53.

Department of Agriculture, but also that of the head of the Office of Drainage Investigation. An attempt to drain the Everglades was made again by a Florida governor from 1926 to 1929. Once more Arthur Morgan, this time in private practice, stepped in to reveal the inadequacy of the plans and thus discourage bond sales.

But schemes affecting the Everglades did not end then. Beginning in 1949, the U.S. Army Corps of Engineers started diverting excess water from the giant Lake Okeechobee to the Gulf of Mexico to reduce the danger of flooding to nearby sugar plantations. As a result, the Everglades, lacking water during the dry season, were drying up. A priceless wildlife refuge was falling prey to humanity's appetite. In addition, the diversion of waters to the Gulf and the ocean also affected human habitations in southern Florida. Cities that once thought they had unlimited supplies of fresh groundwater found they were pumping salt water instead as ocean waters seeped in.[25] Current estimates are that $10 billion will be needed to reverse generations of damage, but initial federal funding faded quickly after the combination of September 11, 2001 and Hurricane Katrina diverted money in other directions.

4. Discuss one of the following topics with an eye to how individual choices in everyday life affect the environment: (a) drinking from disposable cups for coffee or soda pop, (b) driving a sports utility vehicle that gets low gas mileage, (c) eating beef, (d) going the extra mile to dispose of your spent dry cell at a collection point (such as Radio Shack).

5. The social experimentation model of engineering highlights the need to monitor engineering projects after they are put in place. Discuss this idea in connection with Hurricane Katrina.

6. Research the recent approaches, legislation, and international agreements in fighting global warming. Which hold the most promise?

**9.2
Environmental
Moral
Frameworks**

Individual engineers can make a difference. Although their actions are limited—within corporations, they share responsibility with many others—they are uniquely placed to act as agents of change, as responsible experimenters. Doing so requires personal commitments that are often rooted in wider moral or religious frameworks. Here we provide an overview of some of the environmental ethics that are currently being explored, to

---

[25] Arthur E. Morgan, *Dams and Other Disasters*, 370–89.

stimulate further reflection on wider moral frameworks concerning the environment.[26]

Human-Centered Ethics

Human-centered, or anthropocentric, environmental ethics focuses exclusively on the benefits of the natural environment to humans and the threats to human beings presented by the destruction of nature. Each of the ethical theories we examined in Chapter 3—utilitarianism, rights ethics, duty ethics, and virtue ethics—provides a framework for exploring the moral issues concerning the environment. In their classic formulations, all of them assume that, among the creatures on earth, only human beings have inherent moral worth and hence deserve to be taken into account in making moral decisions concerning the environment (or anything else). Other creatures and ecosystems have at most "instrumental value"—as means to promoting human interests.

Utilitarians enjoin us to maximize good consequences for human beings. In developing an environmental ethic, the relevant goods consist of human interests and goods linked to nature. Many of those pleasures and interests concern engineered products made from natural resources. In addition, we have aesthetic interests, as in the beauty of plants, waterfalls, and mountain ranges, and recreational interests, as in hiking and backpacking in wilderness areas. We have scientific interests, especially in the study of "natural labs" of ecological preserves, such as the rain forests. And most basic, we have survival interests, which are linked directly to conserving resources and preserving the natural environment.

The typical argument of rights ethics is that the basic rights to life and to liberty entail a right to a livable environment. The right to a livable environment did not generally enter into people's thinking until the end of the twentieth century, at the time when pollution and resource depletion reached alarming proportions. Nevertheless, it is directly implied by the rights to life and liberty, given that these basic rights cannot be exercised without a supportive natural environment. A right to a livable environment is implied by rights to life and to liberty, and it "imposes upon everyone a correlative moral obligation to respect."[27]

---

[26] Valuable collections of readings include P. Aarne Vesilind and Alastair S. Gunn, *Engineering, Ethics, and the Environment*; and Louis P. Pojman and Paul Pojman, eds., *Environmental Ethics*, 5th ed. (Belmont, CA: Thomson/Wadsworth, 2008).

[27] William T. Blackstone, "Ethics and Ecology," in *Business Ethics*, 2nd ed., ed. W. M. Hoffman and J. M. Moore (New York: McGraw-Hill, 1990), 473.

In duty ethics, which makes duties rather than rights fundamental, respect for human life implies far greater concern for nature than has been traditionally recognized. Kant believed that we owe duties only to rational beings, which in his view excluded all nonhuman animals, although of course he did not have access to recent scientific studies showing striking parallels between humans and other primates. Nevertheless, he condemned callousness and cruelty toward conscious animals because he saw the danger that such attitudes would foster inhumane treatment of persons. In any case, a duty-centered ethics would emphasize the need for conserving the environment because doing so is implied by respect for human beings who depend on it for their very existence.

Finally, virtue ethics draws attention to such virtues as prudence, humility, appreciation of beauty, and gratitude toward the natural world that makes life possible, and also the virtue of stewardship over resources that are needed for further generations. Thomas E. Hill, Jr., offers an anecdote: "A wealthy eccentric bought a house in a neighborhood I know. The house was surrounded by a beautiful display of grass, plants, and flowers, and it was shaded by a huge old avocado tree. But the grass required cutting, the flowers needed tending, and the man wanted more sun. So he cut the whole lot down and covered the yard with asphalt."[28]

The man's attitudes, suggests Hill, are comparable to the callousness shown in strip mining, the cutting of redwood forests, and other destruction of ecosystems with blinkered visions of usefulness.

All these human-centered ethics permit and indeed require a long-term view of conserving the environment, especially because the human beings who have inherent worth will include future generations. Not everything of importance within a human-centered ethics fits neatly into cost-benefit analyses with limited time horizons; much must be accounted for by means of constraints or limits that cannot necessarily be assigned dollar signs. Yet, some have argued that all versions of human-centered ethics are flawed and that we should widen the circle of things that have inherent worth, that is, value in themselves, independent of human desires and appraisals. Especially since 1979, when the journal *Environmental Ethics* was founded, philosophers have explored a wide range of *nature-centered ethics* that, for example, affirm the inherent worth of all conscious animals,

---

[28] Thomas E. Hill, Jr., "Ideals of Human Excellence and Preserving Natural Environments," in *Environmental Virtue Ethics,* ed. Ronald Sandler and Philip Cafaro (Lanham, MD: Rowman and Littlefield, 2005), 47.

of all living organisms, or of ecosystems. Let us consider each of these approaches.

Sentient-Centered Ethics

One version of nature-centered ethics recognizes all sentient animals as having inherent worth. Sentient animals are those that feel pain and pleasure and have desires. Thus, some utilitarians extend their theory (that right action maximizes goodness for all affected) to sentient animals as well as humans. Most notably, Peter Singer developed a revised act-utilitarian perspective in his influential book, *Animal Liberation*. Singer insists that moral judgments must take into account the effects of our actions on sentient animals. Failure to do so is a form of discrimination akin to racism and sexism. He labels it "speciesism": "a prejudice or attitude of bias toward the interests of members of one's own species and against those of members of other species."[29] In Singer's view, animals deserve equal consideration, in that their interests should be weighed fairly, but that does not mean equal treatment with humans (as their interests are different from human interests). Thus, in building a dam that will cause flooding to grasslands, engineers should take into account the impact on animals living there. Singer allows that sometimes animals' interests have to give way to human interests, but their interests should always be considered and weighed.

Singer does not ascribe rights to animals, and hence it is somewhat ironic that *Animal Liberation* has been called the bible of the animal rights movement. Other philosophers, however, do ascribe rights to animals. Most notably, Tom Regan contends that conscious creatures have inherent worth not only because they can feel pleasure and pain, but because more generally they are subjects of experiences who form beliefs, memories, intentions, and preferences, and they can act purposefully.[30] In his view, their status as subjects of experiments makes them sufficiently like humans to give them rights.

Singer and Regan tend to think of inherent worth as all-or-nothing. Hence they think of conscious animals as deserving equal consideration. That does not mean they must be treated in the identical way we treat humans, but only that their interests should be weighed equally with human interests in making decisions. Other sentient-ethicists disagree. They regard conscious

---

[29] Peter Singer, *Animal Liberation*, rev. ed. (New York: Avon Books, 1990), 6.
[30] Tom Regan, *The Case for Animal Rights* (Berkeley, CA: University of California Press, 1983).

animals as having inherent worth, although not equal to that of humans.[31]

Biocentric Ethics

A life-centered ethics regards all living organisms as having inherent worth. Albert Schweitzer (1875–1965) set forth a pioneering version of this perspective under the name of "reverence for life." He argued that our most fundamental feature is not our intellect but instead our will to live, by which he meant both a will to survive and a will to develop according to our innate tendencies. All organisms share these instinctive tendencies to survive and develop, and hence consistency requires that we affirm the inherent worth of all life. More than an appeal to logical consistency, however, Schweitzer appealed to what has been called "bioempathy"—our capacity to experience a kinship with other life, to experience other life in its struggle to survive and grow. Empathy, if we allow it to emerge, grows into sympathy and compassion, gradually leading us to accept "as good preserving life, promoting life, developing all life that is capable of development to its highest possible value."[32]

Schweitzer often spoke of reverence for life as the fundamental excellence of character, and hence his view is a version of nature-centered virtue ethics. He refused to rank forms of life according to degrees of inherent worth, but he believed that a sincere effort to live by the ideal and virtue of reverence for life would enable us to make inevitable decisions about when life must be maintained or has to be sacrificed. More recent defenders of biocentric ethics, however, have developed complex sets of rules for guiding decisions.

Paul Taylor, for example, provides extensive discussion of four duties: (1) nonmaleficence, which is the duty not to kill other living things; (2) noninterference, which is the duty not to interfere with the freedom of living organisms; (3) fidelity, which is the duty not to violate the trust of wild animals (as in trapping); and (4) restitution, which is the duty to make amends for violating the previous three duties.[33] These are prima facie duties, which have exceptions when they conflict with overriding moral duties and rights, such as self-defense.

---

[31] Mary Midgley, *Animals and Why They Matter* (Athens, GA: University of Georgia Press, 1984).

[32] Albert Schweitzer, *Out of My Life and Thought,* trans. A. B. Lemke (New York: Henry Holt and Company, 1990), 157. See also Mike W. Martin, *Albert Schweitzer's Reverence for Life* (Hampshire, UK: Ashgate Publishing, 2007).

[33] Paul W. Taylor, *Respect for Nature* (Princeton, NJ: Princeton University Press, 1986).

Ecocentric Ethics

A frequent criticism of sentient-centered and biocentered ethics is that they are too individualistic, in that they locate inherent worth in individual organisms. Can we seriously believe that each microbe and weed has inherent worth? By contrast, ecocentered ethics locates inherent value in ecological systems. This more holistic approach was voiced by the naturalist Aldo Leopold (1887–1948), who urged that we have an obligation to promote the health of ecosystems. In one of the most famous statements in the environmental literature, he wrote: "A thing is right when it tends to preserve the integrity, stability, and beauty of the biotic community. It is wrong when it tends otherwise."[34] This "land ethic," as he called it, implied a direct moral imperative to preserve (leave unchanged), not just conserve (use prudently), the environment, and to live with a sense that we are part of nature, rather than that nature is a mere resource for satisfying our desires.

More recent defenders of ecocentric ethics have included within this holistic perspective an appreciation of human relationships. Thus, J. Baird Callicott writes that an ecocentric ethic does not "replace or cancel previous socially generated human-oriented duties—to family and family members, to neighbors and neighborhood, to all human beings and humanity."[35] That is, locating inherent worth in wider ecological systems does not cancel out or make less important what we owe to human beings.

Religious Perspectives

Each world religion reflects the diversity of outlooks of its members, and the same is true concerning environmental attitudes. Moreover, these religions have endured through millennia in which shifting attitudes have led a mixed legacy of concern and callousness, with large gaps between ideals and practice. Nevertheless, the potential for world religions to advance ecological understanding is enormous, and we briefly take note of several examples.

Judeo-Christian traditions begin with two contrasting images in *Genesis*. The first chapter portrays God as commanding human dominion over the earth: "Be fruitful and multiply, and fill the earth and subdue it; and have dominion over the fish of the sea and over the birds of the air, and over every living thing that moves on the earth." The second chapter commands "stewardship over all the earth," suggesting the role of caretaker. In principle

---

[34] Aldo Leopold, *A Sand County Almanac* (New York: Ballantine, 1970), 262.
[35] J. Baird Callicott, "Environmental Ethics," in *Encyclopedia of Ethics*, vol. 1, ed. L. C. Becker (New York: Garland, 1992), 313–14.

the two roles are compatible and mutually limiting, especially if "dominion" is interpreted to mean stewardship rather than dominance. In practice, the message of dominance has predominated throughout most of human history in sanctioning unbridled exploitation.[36] Islam also contains a mixed heritage on the environment, with the Koran containing passages that alternate between themes of exploitation of nature for human pleasure and themes of responsible stewardship over what ultimately remains the property of God, not humans.[37]

Today, many concerned Christians, Jews, and Muslims are rethinking their traditions in light of what we have learned. For example, Ian Barbour, a physicist and ecumenical religious thinker, urges that we keep before us the lunar astronauts' pictures of the earth as "a spinning globe of incredible richness and beauty, a blue and white gem among the barren planets," while at the same time exploring "its natural environments and its social order" as we seek together "a more just, participatory, and sustainable society on planet earth."[38]

Asian religions emphasize images of unity with nature, which is distinct from both stewardship and domination. Zen Buddhism, flourishing in Japan, stresses unity of self with nature through immediate, meditative experience. It calls for a life of simplicity and compassion toward the suffering of humans and other creatures. Taoism, rooted in Chinese thought, also accents themes of unity with nature and the universe. The Tao (the Way) is the way of harmony attained by experiencing ourselves as being at one with nature. And Hinduism, the predominant religion in India (powerfully represented by Mahatma Gandhi), promulgates an ideal of oneness with nature and the doctrine of *ahimsa*—nonviolence and nonkilling. It also portrays the sacred and the natural as fused, symbolized in the idea of divinities being reincarnated in living creatures.

Themes of unity are familiar in nineteenth-century English Romanticism and American Transcendentalism. The most deeply rooted American themes of unity, however, are found in American Indian thinking and rituals. Nonhuman animals have spirits. They are to be killed only out of necessity, and then atoned for and apologies made to the animal's spirit. In addition, the identity of tribes was linked to features of the landscape. Unity was

---

[36] Lynn White, "The Historical Roots of Our Ecological Crisis," *Science* 155 (March 10, 1967): 1203–7.
[37] Mawil Y. Izzi Deen (Samarrai), "Islamic Environmental Ethics, Law and Society," in *Ethics of Environment and Development,* ed. J. R. Engel and J. G. Engel (Bellhaven Press, London, 1990).
[38] Ian G. Barbour, *Technology, Environment, and Human Values* (New York: Praeger, 1980), 316.

understood in terms of interdependence and kinship among types of creatures and natural systems.[39]

Many additional approaches could be cited, including forms of spirituality not tied exclusively to particular world religions. For example, feminist outlooks—"ecofeminism"—might or might not be tied to specific religions. They draw parallels between traditional attitudes of dominance and exploitation of men over women and humans over nature. Many, although not all, build on an "ethics of care" that emphasizes themes of personal responsibility, relationships, and contextual reasoning.[40]

We have set forth these environmental ethics in connection with the reflections of individuals, not engineering corporations. Engineering would shut down if it had to grapple with theoretical disputes about human- and nature-centered ethics. Fortunately, at the level of practical issues the ethical theories often converge in the general direction for action, if not in all specifics. Just as humanity is part of nature, human-centered and nature-centered ethics overlap extensively in many of their practical implications.[41] Thus, nature-centered ethics will share with human-centered ethics the justification of human beings' rights to survive, defend themselves, and pursue their self-fulfillment in reasonable ways. Just as it is important for individuals to explore their personal beliefs on this topic, it is equally important for them to seek out and build on areas of overlap, so as to participate in developing responsible social policies and projects.

To conclude, the environment is no longer the concern of an isolated minority. Engineers, corporations, federal and state laws, local community regulations, market mechanisms, and social activists are among the many influences at work. Given the complexity of the issues, we can expect controversy among viewpoints, and nowhere is there a greater need for ongoing dialogue and mutual respect. There is no longer any doubt, however, about the urgency and importance of the issues confronting all of us.

Discussion Questions

1. What ethical theory would you apply to our relation to the environment? Explain why you favor it, and also identify how extensively its practical implications differ from at least two alternative perspectives, selected from those discussed in this section.

---

[39] John Neihardt, *Black Elk Speaks* (Lincoln: University of Nebraska Press, 1961).

[40] Karen J. Warren, ed., *Ecological Feminist Philosophies* (Bloomington, IN: Indiana University Press, 1996).

[41] James P. Sterba, "Reconciling Anthropocentric and Nonanthropocentric Environmental Ethics," in *Ethics in Practice,* ed. Hugh LaFollette (New York: Blackwell, 1997), 644–56.

2. Do you agree or disagree, and why, with Peter Singer's claim that it is a form of bigotry—"speciesism"—to give preference to human interests over the interests of other sentient creatures? Also, should we follow Albert Schweitzer in refusing to rank life forms in terms of their importance?

3. Exxon's 987-foot tanker *Valdez* was passing through Prince William Sound on March 24, 1989, carrying 50 million gallons of oil when it fetched up on Bligh Reef, tore its bottom, and spilled 11 million gallons of oil at the rate of a thousand gallons a second.[42] The immediate cause of the disaster was negligence by the ship's captain, Joseph J. Hazelwood, who was too drunk to perform his duties. Additional procedural violations, lack of emergency preparedness, and a single- rather than double-hull on the ship all contributed in making matters worse. This was one of the worst spills ever, not in quantity, but in its effect on a very fragile ecosystem. No human life was lost, but many thousands of birds, fish, sea otters, and other creatures died.

   Discuss how each of the human-centered and nature-centered ethical theories would interpret the moral issues involved in this case, and apply your own environmental ethic to the case.

4. Discuss the "last person scenario": You are the last person left on earth and can press a button (connected to nuclear bombs) destroying all life on the planet.[43] Is there a moral obligation not to press the button, and why? How would each of the environmental ethics answer this question?

5. Evaluate the following argument from W. Michael Hoffman. In most cases, what is in the best interests of human beings may also be in the best interests of the rest of nature. . . . But if the environmental movement relies only on arguments based on human interests, then it perpetuates the danger of making environmental policy and law on the basis of our strong inclination to fulfill our immediate self-interests. . . . Without some grounding in a deeper environmental ethic with obligations to nonhuman natural things, then the temptation to view our own interests in disastrously short-term ways is that much more encouraged.[44]

---

[42] Art Davidson, *In the Wake of the Exxon Valdez* (San Francisco: Sierra Club Books, 1990).

[43] Richard Routley and Val Routley, "Human Chauvinism and Environmental Ethics," *Environmental Philosophy,* Monograph Series, no. 2, ed. Don Mannison, Michael McRobbie, and Richard Routley (Australian National University, 1980), 121.

[44] W. Michael Hoffman, "Business and Environmental Ethics," in *Business Ethics: Readings and Cases in Corporate Morality,* 4th ed., ed. W. Michael Hoffman, Robert E. Frederick, and Mark S. Schwartz (Boston: McGraw-Hill, 2001), 441.

6. Buckminster Fuller compared the earth to a spaceship. Compare and contrast the moral implications of that analogy with the Gaia Hypothesis set forth by James Lovelock in the passage that follows: "We have . . . defined Gaia as a complex entity involving the Earth's biosphere, atmosphere, oceans, and soil; the totality constituting a feedback or cybernetic system which seeks an optimal physical and chemical environment for life on this planet. The maintenance of relatively constant conditions by active control may be conveniently described by the term 'homoeostasis.'"[45]

What are the strengths and weaknesses of each analogy?

7. Write an essay on one of the following topics: "Why Save Endangered Species?" "Why Save the Everglades?" "What are corporations' responsibilities concerning the environment?" In your essay, explain and apply your environmental ethics.

---

[45] James Lovelock, *Gaia: A New Look at Life on Earth* (New York: Oxford University Press, 2000), 10.

CHAPTER 10

Global Justice

On September 11, 2001, at 8:46 a.m., Al-Qaeda terrorists flew a hijacked American Airlines Boeing 767 into floors 94 to 98 of the 110-story North Tower of the World Trade Center. Seventeen minutes later, the World Trade Center was hit again as more terrorists flew a United Airlines Boeing 767 into floors 78 to 84 of its South Tower. The impact of the airplanes did not collapse the twin towers, but the firestorm set off by the full loads of jet fuel, together with the tons of combustible office material, created intense heat that weakened the steel supports. First the floor trusses weakened and began to tear away from the exterior and interior steel columns, and then the compromised columns gave way. Once the top floors collapsed, it took only 12 seconds for the pancake-like cascade of the South Tower to occur, followed a short time later by the North Tower. A third hijacked airliner was flown into the Pentagon, and a fourth crashed outside Shanksville, Pennsylvania, as its passengers fought the hijackers. The overall death toll was near 3,000 people, including hundreds of firefighters and police officers.

Engineers were prescient in designing the twin towers to withstand impacts from jumbo jets, but they only envisioned jets that were moving slowly and, with depleted fuel, making emergency landings; they had not imagined the possibility of a terrorist attack like the one on September 11, nor had anyone else (Figure 10–1). Because airplanes had crashed into tall buildings before, engineer James Sutherland had warned in 1974 of the vulnerability of hundreds of skyscrapers to further crashes, but it was only in 1994 that the warning was taken seriously after a terrorist plot to hijack an Algerian airliner to attack Paris was foiled.[1] In addition, a bold decision (costing $300,000) was made

---

[1] R. J. M. Sutherland, "The Sequel to Ronan Point," *Proceedings, 42nd Annual Convention, Structural Engineers Association of California* (October 4–6, 1973), 167, See also James R. Chiles, *Inviting Disaster: Lessons from the Edge of Technology* (New York: HarperBusiness, 2002).

Figure 10–1
World Trade Center

©2000 Image 100 Ltd.

during construction of the towers to replace asbestos insulation, which had health dangers that were only then becoming clear and which had already been used on the first 34 floors of the towers, with new fireproofing material coming on the market.[2] The impact of the crash, however, stripped the insulation from the steel beams, leaving them unprotected from temperatures over 1,100°F. Nor was there a safe exit for people above the impact area, as sprinkler systems, emergency elevators, and stairways were damaged by the crash. Fortunately, the buildings stood long enough for some 25,000 people to escape.

As the tallest buildings in New York City and as centers of international commerce, the Twin Towers symbolized the global economy and America's dominance within that economy. The terrorists were fanatics who opposed Western capitalism, democracy, and moral pluralism. Politicians portrayed the attack as an assault on civilization, but perhaps a more accurate statement is that the violence expressed "tensions built into a single global

---

[2] Angus Kress Gillespie, *Twin Towers: The Life of New York City's World Trade Center* (New York: New American Library, 2002), 117–18.

civilization as it emerges against a backdrop of traditional ethnic and religious divisions."[3]

Globalization refers to the increasing integration of nations through trade, investment, transfer of technology, and exchange of ideas and culture. Daniel Yergin and Joseph Stanislaw distinguish a narrow and broader sense of *globalization*: "In a more narrow sense, it represents an accelerating integration and interweaving of national economies through the growing flows of trade, investment, and capital across historical borders. More broadly, those flows include technology, skills and culture, ideas, news, information, and entertainment—and, of course, people. Globalization has also come to involve the increasing coordination of trade, fiscal, and monetary policies among countries."[4]

Today's interdependence among societies—economic, political, and cultural—is unprecedented in its range and depth. So are the possibilities for increased unity and increased fractures during the process of globalization. Global interdependency affects engineering and engineers in many ways, including the Internet issues addressed in Chapter 8 and the environmental issues discussed in Chapter 9. In this chapter we discuss issues concerning multinational corporations and military work.

10.1 Multinational Corporations

Multinational corporations conduct extensive business in more than one country. In some cases, their operations are spread so thinly around the world that their official headquarters in any one *home country*, as distinct from the additional *host countries* in which they do business, is largely incidental and essentially a matter of historical circumstance or of selection based on tax advantages. The benefits to U.S. companies of doing business in less economically developed countries are clear: inexpensive labor, availability of natural resources, favorable tax arrangements, and fresh markets for products. The benefits to the participants in developing countries are equally clear: new jobs, jobs with higher pay and greater challenge, transfer of advanced technology, and an array of social benefits from sharing wealth.

Yet moral challenges arise, accompanying business and social complications. Who loses jobs at home when manufacturing is taken "offshore"? What does the host country lose in resources, control over its own trade, and political independence? And what are the moral responsibilities of corporations and individuals

---

[3] Benjamin R. Barber, "Democracy and Terror in the Era of Jihad vs. McWorld," in *Worlds in Collision: Terror and the Future of Global Order*, ed. Ken Booth and Tim Dunne (New York: Palgrave Macmillan, 2002), 249.

[4] Daniel Yergin and Joseph Stanislaw, *The Commanding Heights: The Battle for the World Economy* (New York: Touchstone, 2002), 383. See also Joseph E. Stiglitz, *Making Globalization Work* (New York: W.W. Norton, 2006).

operating in less economically developed countries? Here we focus on the last question. Before doing so it will be helpful to introduce the concepts of technology transfer and appropriate technology.

Technology Transfer and Appropriate Technology

Technology transfer is the process of moving technology to a novel setting and implementing it there. Technology includes both hardware (machines and installations) and technique (technical, organizational, and managerial skills and procedures). A novel setting is any situation containing at least one new variable relevant to the success or failure of a given technology. The setting may be within a country where the technology is already used elsewhere, or a foreign country, which is our present interest. A variety of agents may conduct the transfer of technology: governments, universities, private volunteer organizations (such as Engineers Without Borders), consulting firms, and multinational corporations.

In most instances, the transfer of technology from a familiar to a new environment is a complex process. The technology being transferred may be one that originally evolved over a period of time and is now being introduced as a ready-made, completely new entity into a different setting. Discerning how the new setting differs from familiar contexts requires the imaginative and cautious vision of "cross-cultural social experimenters."

The expression *appropriate technology* is widely used, but with a variety of meanings. We use it in a generic sense to refer to identification, transfer, and implementation of the most suitable technology for a new set of conditions. Typically the conditions include social factors that go beyond routine economic and technical engineering constraints. Identifying them requires attention to an array of human values and needs that may influence how a technology affects the novel situation. Thus, in the words of Peter Heller, "appropriateness may be scrutinized in terms of scale, technical and managerial skills, materials/energy (assured availability of supply at reasonable cost), physical environment (temperature, humidity, atmosphere, salinity, water availability, etc.), capital opportunity costs (to be commensurate with benefits), but especially human values (acceptability of the end-product by the intended users in light of their institutions, traditions, beliefs, taboos, and what they consider the good life)."[5]

Examples include the introduction of agricultural machines and long-distance telephones. A country with many poor farmers

---

[5] Peter B. Heller, *Technology Transfer and Human Values* (New York: University Press of America, 1985), 119.

can make better immediate use of small, single- or two-wheeled tractors that can serve as motorized ploughs, to pull wagons or to drive pumps, than it can of huge diesel tractors that require collectivized or agribusiness-style farming. Conversely, the same country can benefit more from the latest in wireless communication technology to spread its telephone service to more people and over long distances than it can from old-fashioned transmission by wire.

Appropriate technology also implies that the technology should contribute to and not detract from *sustainable* development of the host country by providing for careful stewardship of its natural resources and not degrading the environment beyond its carrying capacity. Nor should technology be used to replace large numbers of individually tended small fields by large plantations to grow crops for export, leaving most of the erstwhile farmers jobless and without a source of home grown food.

The word *appropriate* is vague until we answer the questions, appropriate to what, and in what way?[6] Answering those questions immediately invokes values about human needs and environmental protection, as well as facts about situations, making it obvious that *appropriate* is a value-laden term. In this broader sense, the appropriate technology might sometimes be small-, intermediate-, or large-scale technology. Appropriate technology is a generic concept that applies to all attempts to emphasize wider social factors when transferring technologies. As such, it reinforces and amplifies our view of engineering as social experimentation.

With these distinctions in mind, let us turn to a classic case study illustrating the complexities of engineering within multinational settings.

Bhopal

Union Carbide in 1984 operated in 37 host countries in addition to its home country, the United States, ranking 35th in size among U.S. corporations. On December 3, 1984, the operators of Union Carbide's plant in Bhopal, India, became alarmed by a leak and overheating in a storage tank. The tank contained methyl isocyanate (MIC), a toxic ingredient used in pesticides. As a concentrated gas, MIC burns any moist part of bodies with which it comes in contact, scalding throats and nasal passages, blinding eyes, and destroying lungs. Within an hour the leak exploded in a gush that sent 40 tons of deadly gas into the atmosphere.[7] The

---

[6] Langdon Winner, *The Whale and the Reactor* (Chicago: University of Chicago Press, 1986), 62.

[7] Paul Shrivastava, *Bhopal, Anatomy of a Crisis* (Cambridge, MA: Ballinger, 1987).

result was the worst industrial accident in history: 500,000 persons exposed to the gas, 2,500 to 3,000 deaths within a few days, 10,000 permanently disabled, 100,000 to 200,000 others injured (the exact figures will always be disputed).

The government of India required the Bhopal plant to be operated entirely by Indian workers. Hence Union Carbide at first took admirable care in training plant personnel, flying them to its West Virginia plant for intensive training. It also had teams of U.S. engineers make regular on-site safety inspections. But in 1982, financial pressures led Union Carbide to relinquish its supervision of safety at the plant, although it retained general financial and technical control. The last inspection by a team of U.S. engineers occurred that year, despite the fact that the team had warned of many of the hazards that contributed to the disaster.

During the following two years safety practices eroded. One source of the erosion was personnel: high turnover of employees, failure to properly train new employees, and low technical preparedness of the local labor pool. The other source was the move away from U.S. standards (contrary to Carbide's written policies) toward lower Indian standards. By December 1984, there were several extreme hazards, including overloading of the tanks storing the MIC gas and lack of proper cooling of the tanks.

According to the official account, a disgruntled employee unscrewed a pressure gauge on a storage tank and inserted a hose into it. He knew and intended that the water he poured into the tank would do damage but did not know it would cause such immense damage. According to another account, a relatively new worker had been instructed by a new supervisor to flush out some pipes and filters connected to the chemical storage tanks. Apparently the worker properly closed valves to isolate the tanks from the pipes and filters being washed, but he failed to insert the required safety disks to back up the valves in case they leaked.

Either way, by the time the workers noticed a gauge showing the mounting pressure and began to feel the sting of leaking gas, they found their main emergency procedures unavailable. The primary defense against gas leaks was a vent-gas scrubber designed to neutralize the gas. It was shut down (and was turned on too late to help), because it was assumed to be unnecessary during times when production was suspended. The second line of defense was a flare tower that would burn off escaping gas missed by the scrubber. It was inoperable because a section of the pipe connecting it to the tank was being repaired. Finally, workers tried to minimize damage by spraying water 100 feet into the air. The gas, however, was escaping from a stack 120 feet high.

Within two hours most of the chemicals in the tank had escaped to form a deadly cloud over hundreds of thousands of people in

Bhopal. As was common in India, desperately poor migrant laborers had become squatters—by the tens of thousands—in the vacant areas surrounding the plant. They had come with hopes of finding any form of employment, as well as to take advantage of whatever water and electricity was available. Virtually none of the squatters had been officially informed by Union Carbide or the Indian government of the danger posed by the chemicals being produced next door to them. The scope of the disaster was greatly increased because of total unpreparedness.

"When in Rome"

What, in general, are the moral responsibilities of multinational corporations, such as Union Carbide and General Motors, and their engineers? One tempting view is that corporations and employees are merely obligated to obey the laws and dominant customs of the host country: "When in Rome do as the Romans do." This view is a version of ethical relativism, discussed in Chapter 2, the claim that actions are morally right within a particular society when (and only because) they are approved by law, custom, or other conventions of that society. Ethical relativism, however, is false because it might excuse moral horrors. For example, it would justify horrendously low safety standards, if that were all a country required. Laws and conventions are not morally self-certifying. Instead, they are always open to criticism in light of moral reasons concerning human rights, the public good, duties to respect people, and virtues.

An opposite view would have corporations and engineers retain precisely the same practices endorsed at home, never making any adjustments to a new culture. This view is a version of *ethical absolutism,* the claim that moral principles have no justified exceptions and that what is morally true in one situation is true everywhere else. Absolutism is false because it fails to take account of how moral principles can come into conflict, forcing some justified exceptions. Absolutism also fails to take account of the many variable facts.

These considerations lead us to endorse *ethical relationalism* (or contextualism): Moral judgments are and should be made in relation to factors that vary from situation to situation, usually making it impossible to formulate rules that are both simple and absolute. Moral judgments are contextual in that they are made in relation to a wide variety of factors—including the customs of other cultures. Note that relationalism only says that foreign customs are morally relevant. It does not say they are automatically decisive or self-authoritative in determining what should be done. This crucial difference sets it apart from ethical relativism.

Relationalism, we should add, is also consistent with *ethical pluralism,* the view that there is more than one justifiable moral perspective. In particular, there may be a number of morally permissible variations in formulating, interpreting, and applying basic moral principles. Not all rational and morally concerned people must see all specific moral issues the same way. This is as true in thinking about multinational corporations as it is in more everyday issues where we recognize that reasonable people can see moral issues differently and still be reasonable. Some version of ethical pluralism is in our view essential in affirming cultural diversity and respecting legitimate differences among individuals and groups.

International Rights

If moral values are open to alternative interpretations, are there nevertheless some minimal standards that must be met? Let us respond to this question within the framework of rights ethics, which is the most commonly applied ethical theory in making cross-cultural moral judgments. A human right, by definition, is a moral entitlement that places obligations on other people to treat one with dignity and respect. If it makes sense at all, it makes sense across cultures, thereby providing a standard of minimally decent conduct that corporations and engineers are morally required to meet.

How can this general doctrine of human rights be applied practically, to help us understand the responsibilities of corporations doing business in other countries? Thomas Donaldson formulates a list of "international rights," human rights that are implied by, but more specific than, the most abstract human rights to liberty and fairness. These international rights have great importance and are often put at risk. Their exact requirements must be understood contextually, depending on the traditions and economic resources available in particular societies. Just as "ought implies can," rights do not require the impossible, and they also apply only within structured societies that provide a framework for understanding how to fairly distribute the burdens associated with them.

Donaldson suggests there are 10 such international rights.[8]

1. The right to freedom of physical movement.
2. The right to ownership of property.
3. The right to freedom from torture.
4. The right to a fair trial.

---

[8] Thomas Donaldson, *The Ethics of International Business* (New York: Oxford University Press, 1989), 81.

5. The right to nondiscriminatory treatment (freedom from discrimination on the basis of such characteristics as race or sex).
6. The right to physical security.
7. The right to freedom of speech and association.
8. The right to minimal education.
9. The right to political participation.
10. The right to subsistence.

These are human rights; as such they place restrictions on how multinational corporations can act in other societies, even when those societies do not recognize the rights in their laws and customs. For example, the right to nondiscriminatory treatment would make it wrong for corporations to participate in discrimination against women and racial minorities even though this may be a dominant custom in the host country. Again, the right to physical security requires supplying protective goggles to workers running metal lathes, even when this is not required by the laws of the host country.

Although these rights have many straightforward cross-cultural applications, they nevertheless need to be applied contextually to take into account some features of the economy, laws, and customs of host countries. Not surprisingly, many difficulties and gray areas arise. One type of problem concerns the level of stringency required in matters such as degrees of physical safety at the workplace. Workers in less economically developed countries are often willing to take greater risks than would be acceptable to workers in the United States. Here Donaldson recommends applying a "rational empathy test" to determine if it is morally permissible for a corporation to participate in the practices of the host country: Would citizens of the home country find the practice acceptable if their home country were in circumstances economically similar to those of the host country? For example, in determining whether a certain degree of pollution is acceptable for a U.S. company with a manufacturing plant located in India, the U.S. company would have to decide whether the pollution level would be acceptable under circumstances where the United States had a comparable level of economic development.

A second, quite different, type of problem arises where the practice is not so directly linked to economic factors, as in racial discrimination. Here Donaldson insists that unless one can do business in the country without engaging in practices that violate human rights, then corporations must simply leave and go to other countries.

Finally, the appeal to human rights is the most powerful cross-cultural ethical theory, but it is not the only one. All major ethical theories, including duty ethics and utilitarianism, rely on a conception of the worth and dignity of human beings. It is true that not all societies are based on ideals of individual liberty and the sanctity of life. Nevertheless, a valid cross-cultural ethics need not assume that all societies have come to appreciate all important values. Just as important, at some minimum level, all cultures share some common values, including bans on gratuitous violence and requirements of mutual support, reciprocity, and fairness—if they did not, they would not exist.[9] How these basic values are understood varies considerably, but they provide a basis for meaningful cross-cultural moral dialogue.

Promoting Morally Just Measures

Richard T. De George agrees that multinational corporations should respect the basic rights of people in the countries where they do business, but he requires more, especially when wealthy countries do business in less economically developed countries. In the spirit of utilitarianism, which calls for promoting the most good for the most people, he also requires that the activities of multinational corporations benefit the host countries in which they do business.[10]

More fully, the business activities of multinational corporations must do more overall good than bad, which means helping the country's overall economy and its workers, rather than benefiting a few corrupt leaders in oppressive regimes. Not only must they pay their fair share of taxes, but they must make sure the products they manufacture or distribute are not causing easily preventable harms.

In addition, the overall impact of the business dealings must tend to promote morally just institutions in the society, not increase unjust institutions. At the same time, corporations should respect the laws and culture of the host country, providing they do not violate basic moral rights. Of course, there is a tension between promoting just institutions and respecting local customs. For example, as U.S. business attempts to encourage a move toward treating women equally in the workplace, they may undermine local customs, often supported by religion, about the appropriate roles for women. Good judgment exercised in good faith, rather than abstract principles, is often the only way to address such practical dilemmas.

---

[9] Sissela Bok, *Common Values* (Columbia, MO: University of Missouri Press, 1995), 13–16.

[10] Richard T. De George, "Ethical Dilemmas for Multinational Enterprise: A Philosophical Overview," in *Ethics and the Multinational Enterprise,* ed. W. Michael Hoffman, Ann E. Lange, and David A. Fedo (New York: University Press of America, 1986).

De George calls for a contextual, case-by-case approach in applying principles of human rights and promoting the good of the host country. For example, what is a "fair wage" to workers in very poor countries? If multinational corporations pay exactly the pay rate of the host country, they will be accused of exploiting workers, especially when that rate is below a living wage sufficient for the person to live with dignity as a human being. If they pay well beyond that rate they will be accused of unfairly stealing the most skilled workers in the society, drawing them away from other companies important to the local economy. De George's guideline is to pay a living wage, even when local companies fail to pay such a wage, but otherwise pay only enough to attract competent workers.

As another example, consider the issue of worker safety in companies that manufacture hazardous chemicals. When is it permissible for the United States to transfer dangerous technology such as asbestos production to another country and then simply adopt that country's safety laws? Workers have the right to informed consent. Even if the host country does not recognize that right, corporations are required to inform workers, in language they can understand, of the dangers. That is a necessary, but not sufficient condition. Workers may be so desperate for income to feed their families that they will work under almost any conditions. Corporations must eliminate great risks when they can while still making a reasonable profit. They must also pay workers for the extra risks they undertake. Exactly what this means is a matter of morally good judgment and negotiation.

Discussion Questions

1. Following the disaster at Bhopal, Union Carbide argued that officials at its U.S. corporate headquarters had no knowledge of the violations of Carbide's official safety procedures and standards. This has been challenged as documents were uncovered showing they knew enough to have warranted inquiry on their part, but let us assume they were genuinely ignorant. Would ignorance free them of responsibility?

2. The moving of hazardous technologies, such as the manufacture of asbestos, to less-developed countries is motivated in part by cheaper labor costs, but another factor is that workers are willing to take greater risks. How does Donaldson's view apply to this issue?

   Also, do you agree with Richard De George's view that taking advantage of this willingness need not be unjust exploitation if several conditions are met: (1) Workers are informed of the risks. (2) They are paid more for taking the risks. (3) The company takes some steps to lower the risks, even if not to the level acceptable for U.S. workers.[11]

How would you assess Union Carbide's handling of worker safety? Take into account the remarks of an Indian worker interviewed *after* the disaster. The worker was then able to stand only a few hours each day because of permanent damage to his lungs. During that time he begged in the streets while he awaited his share of the legal compensation from Union Carbide. When asked what he would do if offered work again in the plant knowing what he knew now, he replied: "If it opened again tomorrow I'd be happy to take any job they offered me. I wouldn't hesitate for a minute. I want to work in a factory, any factory. Before 'the gas' [disaster] the Union Carbide plant was the best place in all Bhopal to work."[12]

3. How would you balance respect for diversity with commitments to respect for individual rights in the following two cases?

   a. You are a woman assigned to work in a Middle Eastern country that requires women to wear traditional clothing, but doing so conflicts with your religious faith; or, you are a man who is a member of a team whose members include women who are required to wear traditional clothing. If you decline the assignment, your career advancement might suffer.

   b. Your company is asked to design a more efficient weaving apparatus whose size is quickly adjustable to young children, and you are assigned to the project. You know that the primary market for the apparatus is countries that use child labor.

4. During 1972 and 1973 the president of Lockheed, A. Carl Kotchian, authorized secret payments totaling around $12 million beyond a contract to representatives of Japan's Prime Minister Tanaka. Later revelations of the bribes helped lead to the resignation of Tanaka and also to the Foreign Corrupt Practices Act that forbade such payments by American-based corporations. In 1995, long after Tanaka's death, the agonizingly slow trial and appeals process came to an end as Japan's Supreme Court reaffirmed the guilty verdicts, but so far no one has been jailed, and the case appears to have had little recent impact on business and politics in Japan.

   Mr. Kotchian believed at that time it was the only way to assure sales of Lockheed's TriStar airplanes in a much-needed market.

---

[11] Ibid.

[12] Fergus M. Bordewich, "The Lessons of Bhopal," *Atlantic Monthly* (March 1987): 30–33.

In explaining his actions, Mr. Kotchian cited the following facts:[13] (1) There was no doubt in his mind that the only way to make the sales was to make the payments. (2) No U.S. law at the time forbade the payments. Only later, in 1977, was the Foreign Corrupt Practices Act signed into law, largely based on the Lockheed scandal forbidding the payment of overseas bribes. (3) The payments were financially worthwhile, for they totaled only 3 percent of an expected $430 million income for Lockheed. (4) The sales would prevent Lockheed layoffs, provide new jobs, and thereby benefit workers' families and their communities as well as the stockholders. (5) He himself did not personally initiate any of the payments, which were all requested by Japanese negotiators. (6) To give the TriStar a chance to prove itself in Japan, he felt he had to "follow the functioning system" of Japan. That is, he viewed the secret payments as the accepted practice in Japan's government circles for this type of sale.

Present and defend your view about whether Mr. Kotchian's actions were morally justified. In doing so, apply utilitarianism, rights ethics, and other ethical theories that you see as relevant.

5. The World Trade Organization (WTO) was established to oversee trade agreements, enforce trade rules, and settle disputes. Some troublesome issues have arisen when WTO has denied countries the right to impose environmental restrictions on imports from other countries. Thus, for example, the United States may not impose a ban on fish caught with nets that can endanger other sea life such as turtles or dolphins, while European countries and Japan will not be able to ban imports of beef from U.S. herds injected with antibiotics. Yet, other countries ban crops genetically modified to resist certain pests, or products made therefrom, unless labeled as such. Investigate the current disputes and, using a case study, discuss how such problems may be resolved.[14]

6. Corporations' codes of ethics have to take into account international contexts. Compare and contrast the benefits and liabilities of the types of ethics programs (a and b) at Texas Instruments at two different times, described as follows:

   a. Texas Instruments (TI) always had a long-standing emphasis on trust and integrity, but during the 1980s it greatly intensified

---

[13] Carl Kotchian, "The Payoff: Lockheed's 70-day Mission to Tokyo," *Saturday Review* (July 9, 1977).

[14] A useful starting place is Joseph E. Stiglitz, *Making Globalization Work* (New York: W.W. Norton, 2006).

its efforts to make ethics central to the corporation.[15] In 1987 TI appointed a full-time ethics director, Carl Skooglund, who was then a vice president for the corporation. Skooglund reported to an ethics committee that in turn reported directly to the board of directors. His activities included raising employees' ethical awareness through discussion groups and workshops on ethics, making himself directly available to all employees through a confidential phone line, and—especially relevant here—addressing specific cases and concerns in weekly newsletters and detailed brochures called "Cornerstones."

b. In 1995, TI's popular chairman died suddenly, prompting a rapid review of its policies. In two years, it made 20 acquisitions and divestitures, including selling its defense-industry business, leaving it with more non-U.S. employees than U.S. employees. The new chairman called for rethinking its ethics programs to have both a greater international focus and more emphasis on a competitive and "winning" attitude. Before his retirement, Carl Skooglund scrapped the Cornerstone series, focused on specific issues and cases, and replaced it with three core values: integrity (honesty together with respect and concern for people), innovation, and commitment (take responsibility for one's conduct).

10.2 Weapons Development and Peace

Much of the world's technological activity has a military stimulus. Based just on size of expenditures, direct or indirect involvement of engineers, and startling new developments, military technology would deserve serious discussion in engineering ethics. In addition, involvement in military work raises issues connected with the personal commitments of individual engineers.

Involvement in Weapons Work

Historically, a quick death in battle by sword was considered acceptable, whereas the use of remote weapons (from bow and arrow to firearms) was frequently decried as cowardly, devoid of valor, and tantamount to plain murder.[16] As modern weapons of war progressed through catapults, cannons, machine guns, and bombs released from airplanes and missiles to reach further and further, the soldiers firing them were less likely to see the individual human beings—soldiers as well as civilians—they had as their general target. The continuing automation of the battle scene tends to conceal the horrors of war and thus makes

---

[15] Francis J. Aguilar, *Managing Corporate Ethics* (New York: Oxford University Press, 1994), 120–35, 140–43.

[16] Martin von Creveld, *Technology and War, from 2000 BC to the Present* (New York: The Free Press/Macmillan, 1989), 71.

military activity seem less threatening and high-tech wars more appealing.

How might the men and women who design weapons, manufacture them, and use them feel about their work? For some engineers, involvement in weapons development conflicts with personal conscience; for others, it is an expression of conscientious participation in national defense. The following cases illustrate the kinds of moral issues involved in deciding whether to engage in military work.

1. Bob's employer manufactures antipersonnel bombs. By clustering 665 guava-size bomblets and letting them explode above ground, an area covering the equivalent of 10 football fields is subjected to a shower of sharp fragments. Alternatively, the bombs can be timed to explode hours apart after delivery. Originally the fragments were made of steel, and thus they were often removable with magnets; now plastic materials are sometimes used, making the treatment of wounds, including the location and removal of the fragments, more time-consuming for the surgeon. Recently another innovation was introduced: By coating the bomblets with phosphorus, the fragments could inflict internal burns as well. Thus, the antipersonnel bomb does its job quite well without necessarily killing in that it ties up much of the enemy's resources just in treating the wounded who have survived.

   Bob himself does not handle the bombs in any way, but as an industrial engineer he enables the factory to run efficiently. He does not like to be involved in making weapons, but then he tells himself that someone has to produce them. If he does not do his job, someone else will, so nothing would change. Furthermore, with the cost of living being what it is, he owes his family a steady income.

2. Mary is a chemical engineer. A promotion has gotten her into napalm manufacturing. She knows it is nasty stuff, having heard that the Nobel laureate, Professor Wald of Harvard University, was said to have berated the chemical industry for producing this "most brutal and destructive weapon that has ever been created." She saw a scary old photograph from the Vietnam War period, depicting a badly burned peasant girl running from a village in flames. But the locals were said to take forever in leaving a fighting zone and then there were complaints about them being hurt or killed. She abhors war like most human beings, but she feels that the government knows more than she does about international dangers and that the present use of napalm by U.S. forces in Iraq may be unavoidable. Regarding her own future, Mary knows that if she continues to do well on her job she will be promoted, and one of these days she may well be in

the position to steer the company into the production of peaceful products. Will Mary use a higher position in the way she hopes to do, or will she instead wait until she becomes the CEO?

3. Ron is a specialist in missile control and guidance. He is proud to be able to help his country through his efforts in the defense industry, especially as part of the "war on terrorism." The missiles he works on will carry single or multiple warheads with the kind of dreadful firepower which, in his estimation, has kept any potential enemy in check since 1945. At least there has not been another world war—the result of mutual deterrence, he believes.

4. Marco's foremost love is physical electronics. He works in one of the finest laser laboratories. Some of his colleagues do exciting research in particle beams. That the laboratory is interested in developing something akin to the "death ray" described by science fiction writers of his youth is of secondary importance. More bothersome is the secrecy that prevents him from freely exchanging ideas with experts across the world. But why change jobs if he will never find facilities like those he has now?

5. Joanne is an electronics engineer whose work assignment includes avionics for fighter planes that are mostly sold abroad. She has no qualms about such planes going to what she considers friendly countries, but she draws the line at their sale to potentially hostile nations. Joanne realizes that she has no leverage within the company, so she occasionally alerts journalist friends with news she feels all citizens should have. "Let the voters direct the country at election time"—that is her motto.

6. Ted's background and advanced degrees in engineering physics gave him a ready entry into nuclear bomb development. As a well-informed citizen he is seriously concerned with the dangers of the ever-growing nuclear arsenal. He is also aware of the possibilities of an accidental nuclear exchange. In the meantime he is working hard to reduce the risk of accidents such as the 32 "broken arrows" (incidents when missile launchings may have occurred erroneously) that had been reported by the Pentagon during the height of the Cold War, or the many others that he knows have occurred worldwide. Ted continues in his work because he believes that only specialists, with firsthand experience of what modern weapons can do, can eventually turn around the suicidal trend represented by their development. Who else can engage in meaningful arms control negotiations?

As these examples suggest, prudent self-interest is not sufficient to guarantee responsible participation in what must be regarded as humankind's most crucial engineering experiment. We must rely on individuals who have arrived at morally autonomous, well-reasoned positions for either engaging in or abstaining from weapons work to consistently and carefully monitor the experiment and try to keep it from running a wild course.

Defense Industry Problems

Nations confer special privileges on their defense industries, often without giving sufficient thought to the problems that can accompany large military buildups. Unethical business practices, for instance, occur as in all massive projects, but the urgency of completing a weapons system before it becomes obsolete and the secrecy that surrounds it make proper oversight especially difficult. This is one of the problems we describe briefly in this section. The other problems are even more serious because they often go unrecognized.

The problem of waste and cost overruns is a continuing one in the defense industry.[17] One example is the $2 billion cost overrun on the development of the C5-A transport plane, revealed to the U.S. Senate by Ernest Fitzgerald, a Pentagon specialist in management systems. Fitzgerald has been a critic of how the defense industry has operated at efficiencies far below commercial standards. He has described how contractors' workforces were swelled by underused engineers and high-salary sales personnel, resulting in lavish overhead fees. Or how small contractors were willing to comply with cost-cutting plans, but large suppliers felt secure in not complying.

High cost and poor quality were encouraged in various ways: Planned funding levels were leaked to prospective contractors. Cost estimates were based on historical data, thus incorporating past inefficiencies. Costs were cut when necessary by lowering quality, especially when component specifications were not ready until after the contract was signed. Sole-supplier policies gave a contractor the incentive to "buy in" with an artificially low bid, only to plead for additional funds later on. And those funds were usually forthcoming, because the Department of Defense has historically accepted what it knows to be optimistically low development-cost estimates because they stand a better chance of being approved by Congress.[18]

In Goethe's poem *Der Zauberlehrling,* the sorcerer's apprentice employs his master's magic incantation to make the broom fetch water. When he cannot remember the proper command to stop the helpful broom, however, he comes near to drowning before the master returns. Military technology often resembles the sorcerer's

---

[17] Ernest Fitzgerald, *The High Priests of Waste* (New York: W.W. Norton, 1972); J. Gansler, *The Defense Industry* (Cambridge, MA: MIT Press, 1980).

[18] Gansler, *The Defense Industry*, 296.

broom. Not only has the world's arsenal grown inordinately expensive (even without graft), and not only does it contribute to a steadily worsening inflation, it has also gained a momentum all its own.

The technological imperative—that innovations must be implemented—should even give advocates of preparedness for conventional, limited war some cause for concern. Giving in to the excitement of equipping and trying out weapons employing the latest in technology may provide added capability to sophisticated, fully automatic systems such as intercontinental ballistic missiles. But if tactical, humanly operated weapons fall prey to the gadget craze, a less-than- optimal system may result. The F-15 fighter illustrates this problem of preoccupation with prestige-boosting modernism. The plane was the fastest and most maneuverable of its kind, yet 40 percent of the F-15s were not available for service at any one time because of defects, difficulty of repair, and lack of spare parts.

A further problem concerns peacetime secrecy in work of military import. Secrecy poses problems for engineers in various ways. Should discoveries of military significance always be made available to the government? Can they be shared with other researchers, with other countries? Or should they be withheld from the larger scientific and public community altogether? If governmental secrecy in weapons development is allowed to become all-pervasive, however, will it also serve to mask corruption or embarrassing mistakes within the defense establishment? Can secrecy contribute to the promotion of particular weapon systems, such as the X-ray laser, without fear of criticism?[19] There are no easy answers to these questions, and they deserve to be discussed more widely within the defense establishment and in the public.

Peace Engineering

The threat of terrorism has introduced a new variable into much engineering. Designs of such things as bridges, tall buildings, reservoirs, and energy facilities must now take into account the possibility of intentional assaults on structures and processes. Research and development of many new anti-terrorist technologies has greatly accelerating. Some of these technologies have raised new ethical concerns, especially privacy concerns about using software to monitoring e-mail and phone calls, and filming people as they enter public sports events to apply facial identification technologies.

---

[19] John A. Adam, "Dispute over X-Ray Laser Made Public, Scientists' Criticisms Shunted," *Institute* (February 1988): 1.

Concerns about violence inspire some engineers to make personal commitments to attack the deeper sources of injustice and suffering by engaging in "peace engineering." Daniel Vallero and Aarne Vesilind characterize peace engineering as "the proactive use of engineering skills to promote a peaceful and just existence for all people."[20] Examples include working full-time for the Peace Corps, the World Bank, and Engineers Without Borders, as well as volunteering some of one's time to any number of worthy philanthropic groups. Vallero and Vesilind acknowledge that military work also qualifies as peace engineering, when done in a sincere spirit of defending freedom and working toward goals of peace.

As an example, consider work on demining technology, that is, technology used to remove mines, especially land mines that continue to cause untold suffering. Each year, each day, there is a grim harvest of severed limbs and dead bodies as people and their treasured water buffalos tread on mines planted by all sides in Cambodia and Vietnam in the 1960s and 1970s. Afghanistan, Angola, Bosnia, Mozambique, Nicaragua, and Somalia are other countries infested by such antipersonnel weapons. They are easily spread by air but painstakingly difficult and dangerous to remove. The U.S. State Department estimates that 85 to 100 million land mines still remain scattered in these countries and those that were involved in the two world wars.

Whereas it costs only a few dollars to make most land mines, it costs from $300 to $1,000 to remove one.[21] Technical challenges include not only minimizing the costs of defusing or harmlessly exploding mines once they are identified, but also identifying mines more efficiently by reducing the number of false alarms caused by misidentifying other natural objects when using high-tech ground-penetrating radar. Even today the most widely used technology remains 1940s-style metal detectors, which are sorely lacking in tackling the scale of the problem.

As a second illustration of peace engineering, consider the need for innovative software for logistics in responding to the aftermath of war as well as to humanitarian crises. Hurricane Katrina made clear how even the world's most advanced industrial country was ill-prepared to deal with large-scale devastation. The problem is magnified in responding to ethnic wars and mass starvation in places such as Rwanda and the Sudan.

---

[20] Daniel A. Vallero and P. Aarne Vesilind, *Socially Responsible Engineering* (Hoboken, NJ: John Wiley and Sons, 2007), 124. See also P. Aarne Vesilind, ed., *Peace Engineering: When Personal Values and Engineering Careers Converge* (Woodsville, NH: Lakeshore Press, 2005); and George D. Catalano, *Engineering Ethics: Peace, Justice, and the Earth* (Morgan and Claypool, 2006).

[21] Regina E. Dugan, "Demining Technology," in *Technology for Humanitarian Action*, ed. Kevin M. Cahill (New York: Fordham University Press, 2005), 272.

Everything from international fundraising to air and ground transportation to medical and security personnel needs to be coordinated. For example, a pioneering Web-based Humanitarian Logistics Software, developed by the Fritz Institute, has several modules: "The mobilization module simultaneously tracks needs of the beneficiaries and agency funding appeals, reconciling them with donations. The procurement module controls purchase orders, performs competitive bid analysis and reconciles received goods against invoices awaiting payment. The transportation and tracking module allows consolidation of supplies for transportation and allows the automatic tracking of major milestones in this process."[22] Relevant data on performance is made available to donors and decision makers.

Finally, in addition to responding to the aftermath of war and to humanitarian crises, peace engineering makes its most enduring contribution by finding innovative ways to fight deeper sources of suffering and injustice. Muhammad Yunus, the 2006 Nobel Peace Prize laureate, demonstrated how even one cell phone in a Bangladesh village, made possible by a micro-loan and electricity from solar energy, can advance economic opportunities for the poorest of the poor.[23] As he began to experiment with a similar village connection to the Internet, MIT professor Nicholas Negroponte had an even more daring vision: Put a laptop computer in the hands of every child. He and his colleagues designed a laptop computer that could be marketed at $100. Nicknamed the green machine, the computer is simple enough to be used by children in developing countries, and is made tough enough to endure their harsh conditions. The computer sparked competition that is likely to drive prices down even further.

Engineers can serve peace and justice in myriad ways. As creators of new technologies, innovative entrepreneurs, leaders in business and government, and volunteers in local communities and global philanthropic projects, they can blend technical skills with reasonable awareness of dangers. They are most effective when they combine these resources with moral imagination and insight.

Discussion Questions

1. The following problem is taken from an article by Tekla Perry in the *IEEE Spectrum*. Although it involves the U.S. National

---

[22] Geoffrey W. Clark, Frank L. Fernandez, and Zhen Zhang, "Technology and Humanitarian Actions: A Historical Perspective, in *Technology for Humanitarian Action*, ed. Kevin M. Cahill 18–19.

[23] Muhammad Yunus, *Banker to the Poor: Micro-Lending and the Battle Against World Poverty* (New York: PublicAffairs, 2003), 225.

Aeronautics and Space Administration rather than the Defense Department, many of the actors (companies and government) involved in space research are also involved in weapons development:

> Arthur is chief engineer in a components house. As such, he sits in meetings concerning bidding on contracts. At one such meeting between top company executives and the National Aeronautics and Space Administration, which is interested in getting a major contract, NASA presents specifications for components that are to be several orders of magnitude more reliable than the current state of the art. The components are not part of a life-support system, yet are critical for the success of several planned experiments. Arthur does not believe such reliability can be achieved by his company or any other, and he knows the executives feel the same. Nevertheless, the executives indicate an interest to bid on the contract without questioning the specifications. Arthur discusses the matter privately with the executives and recommends that they review the seemingly technical impossibility with NASA and try to amend the contract. The executives say that they intend, if they win the contract, to argue mid-stream for a change. They remind Arthur that if they don't win the contract, several engineers in Arthur's division will have to be laid off. Arthur is well-liked by his employees and fears the layoffs would affect some close friendships. What should Arthur do?[24]

2. Are there any ethical grounds for maintaining a large nuclear stockpile today? Discuss any stabilizing or destabilizing effects you see.[25]

3. The just-war theory considers a war acceptable when it satisfies several stringent criteria: The war must be fought for a just cause, the motives must be good, it must follow a call from a legitimate authority, and the use of force must be based on necessity (as a last resort).[26] Central to notions of a just war are the principles of noncombatant immunity and proportionality. Noncombatants are those who will not be actively participating in combat and therefore do not need to be killed or restrained. Proportionality addresses the extent of damage or consequences allowable in terms of need and cost. Describe a scenario for the conduct of

---

[24] Tekla S. Perry, "Five Ethical Dilemmas," *IEEE Spectrum* 18 (June 1981): 58. Quotations in text used with permission of the author and the Institute of Electrical and Electronics Engineers.

[25] See Gregory S. Kavka, *Moral Paradoxes of Nuclear Deterrence* (Cambridge: Cambridge University Press, 1987).

[26] Aaron Fichtelberg, "Applying the Rules of Just War Theory to Engineers in the Arms Industry," *Science and Engineering Ethics* 12 (2006): 685–700. See also James E. White, ed., *Contemporary Moral Problems: War and Terrorism*, 2nd ed. (Belmont, CA: Thomson Wadsworth, 2006).

a just war and describe the kinds of weapons engineers might have to develop to wage one. In your view, is the U.S.-Iraq war a just war?

4. Wernher von Braun designed Hitler's V-2 rocket that terrorized London toward the end of World War II; after the War he worked for the U.S. Army and designed the Saturn rockets that launched Apollo astronauts to the moon. Research his career and comment on its implications for thinking about engineering.

**National Society of
Professional Engineers®**

Code of Ethics for Engineers

Preamble

Engineering is an important and learned profession. As members of this profession, engineers are expected to exhibit the highest standards of honesty and integrity. Engineering has a direct and vital impact on the quality of life for all people. Accordingly, the services provided by engineers require honesty, impartiality, fairness, and equity, and must be dedicated to the protection of the public health, safety, and welfare. Engineers must perform under a standard of professional behavior that requires adherence to the highest principles of ethical conduct.

I. Fundamental Canons

Engineers, in the fulfillment of their professional duties, shall:
1. Hold paramount the safety, health, and welfare of the public.
2. Perform services only in areas of their competence.
3. Issue public statements only in an objective and truthful manner.
4. Act for each employer or client as faithful agents or trustees.
5. Avoid deceptive acts.
6. Conduct themselves honorably, responsibly, ethically, and lawfully so as to enhance the honor, reputation, and usefulness of the profession.

II. Rules of Practice

1. Engineers shall hold paramount the safety, health, and welfare of the public.
   a. If engineers' judgment is overruled under circumstances that endanger life or property, they shall notify their employer or client and such other authority as may be appropriate.
   b. Engineers shall approve only those engineering documents that are in conformity with applicable standards.
   c. Engineers shall not reveal facts, data, or information without the prior consent of the client or employer except as authorized or required by law or this Code.
   d. Engineers shall not permit the use of their name or associate in business ventures with any person or firm that they believe is engaged in fraudulent or dishonest enterprise.
   e. Engineers shall not aid or abet the unlawful practice of engineering by a person or firm.
   f. Engineers having knowledge of any alleged violation of this Code shall report thereon to appropriate professional bodies and, when relevant, also to public authorities, and cooperate with the proper authorities in furnishing such information or assistance as may be required.
2. Engineers shall perform services only in the areas of their competence.
   a. Engineers shall undertake assignments only when qualified by education or experience in the specific technical fields involved.
   b. Engineers shall not affix their signatures to any plans or documents dealing with subject matter in which they lack competence, nor to any plan or document not prepared under their direction and control.
   c. Engineers may accept assignments and assume responsibility for coordination of an entire project and sign and seal the engineering documents for the entire project, provided that each technical segment is signed and sealed only by the qualified engineers who prepared the segment.
3. Engineers shall issue public statements only in an objective and truthful manner.
   a. Engineers shall be objective and truthful in professional reports, statements, or testimony. They shall include all relevant and pertinent information in such reports, statements, or testimony, which should bear the date indicating when it was current.
   b. Engineers may express publicly technical opinions that are founded upon knowledge of the facts and competence in the subject matter.
   c. Engineers shall issue no statements, criticisms, or arguments on technical matters that are inspired or paid for by interested parties, unless they have prefaced their comments by explicitly identifying the interested parties on whose behalf they are speaking, and by revealing the existence of any interest the engineers may have in the matters.

4. Engineers shall act for each employer or client as faithful agents or trustees.
   a. Engineers shall disclose all known or potential conflicts of interest that could influence or appear to influence their judgment or the quality of their services.
   b. Engineers shall not accept compensation, financial or otherwise, from more than one party for services on the same project, or for services pertaining to the same project, unless the circumstances are fully disclosed and agreed to by all interested parties.
   c. Engineers shall not solicit or accept financial or other valuable consideration, directly or indirectly, from outside agents in connection with the work for which they are responsible.
   d. Engineers in public service as members, advisors, or employees of a governmental or quasi-governmental body or department shall not participate in decisions with respect to services solicited or provided by them or their organizations in private or public engineering practice.
   e. Engineers shall not solicit or accept a contract from a governmental body on which a principal or officer of their organization serves as a member.
5. Engineers shall avoid deceptive acts.
   a. Engineers shall not falsify their qualifications or permit misrepresentation of their or their associates' qualifications. They shall not misrepresent or exaggerate their responsibility in or for the subject matter of prior assignments. Brochures or other presentations incident to the solicitation of employment shall not misrepresent pertinent facts concerning employers, employees, associates, joint venturers, or past accomplishments.
   b. Engineers shall not offer, give, solicit, or receive, either directly or indirectly, any contribution to influence the award of a contract by public authority, or which may be reasonably construed by the public as having the effect or intent of influencing the awarding of a contract. They shall not offer any gift or other valuable consideration in order to secure work. They shall not pay a commission, percentage, or brokerage fee in order to secure work, except to a bona fide employee or bona fide established commercial or marketing agencies retained by them.

III. Professional Obligations

1. Engineers shall be guided in all their relations by the highest standards of honesty and integrity.
   a. Engineers shall acknowledge their errors and shall not distort or alter the facts.
   b. Engineers shall advise their clients or employers when they believe a project will not be successful.
   c. Engineers shall not accept outside employment to the detriment of their regular work or interest. Before accepting any outside engineering employment, they will notify their employers.
   d. Engineers shall not attempt to attract an engineer from another employer by false or misleading pretenses.
   e. Engineers shall not promote their own interest at the expense of the dignity and integrity of the profession.
2. Engineers shall at all times strive to serve the public interest.
   a. Engineers are encouraged to participate in civic affairs; career guidance for youths; and work for the advancement of the safety, health, and well-being of their community.
   b. Engineers shall not complete, sign, or seal plans and/or specifications that are not in conformity with applicable engineering standards. If the client or employer insists on such unprofessional conduct, they shall notify the proper authorities and withdraw from further service on the project.
   c. Engineers are encouraged to extend public knowledge and appreciation of engineering and its achievements.
   d. Engineers are encouraged to adhere to the principles of sustainable development[1] in order to protect the environment for future generations.

3. Engineers shall avoid all conduct or practice that deceives the public.
   a. Engineers shall avoid the use of statements containing a material misrepresentation of fact or omitting a material fact.
   b. Consistent with the foregoing, engineers may advertise for recruitment of personnel.
   c. Consistent with the foregoing, engineers may prepare articles for the lay or technical press, but such articles shall not imply credit to the author for work performed by others.
4. Engineers shall not disclose, without consent, confidential information concerning the business affairs or technical processes of any present or former client or employer, or public body on which they serve.
   a. Engineers shall not, without the consent of all interested parties, promote or arrange for new employment or practice in connection with a specific project for which the engineer has gained particular and specialized knowledge.
   b. Engineers shall not, without the consent of all interested parties, participate in or represent an adversary interest in connection with a specific project or proceeding in which the engineer has gained particular specialized knowledge on behalf of a former client or employer.
5. Engineers shall not be influenced in their professional duties by conflicting interests.
   a. Engineers shall not accept financial or other considerations, including free engineering designs, from material or equipment suppliers for specifying their product.
   b. Engineers shall not accept commissions or allowances, directly or indirectly, from contractors or other parties dealing with clients or employers of the engineer in connection with work for which the engineer is responsible.
6. Engineers shall not attempt to obtain employment or advancement or professional engagements by untruthfully criticizing other engineers, or by other improper or questionable methods.
   a. Engineers shall not request, propose, or accept a commission on a contingent basis under circumstances in which their judgment may be compromised.
   b. Engineers in salaried positions shall accept part-time engineering work only to the extent consistent with policies of the employer and in accordance with ethical considerations.
   c. Engineers shall not, without consent, use equipment, supplies, laboratory, or office facilities of an employer to carry on outside private practice.
7. Engineers shall not attempt to injure, maliciously or falsely, directly or indirectly, the professional reputation, prospects, practice, or employment of other engineers. Engineers who believe others are guilty of unethical or illegal practice shall present such information to the proper authority for action.
   a. Engineers in private practice shall not review the work of another engineer for the same client, except with the knowledge of such engineer, or unless the connection of such engineer with the work has been terminated.
   b. Engineers in governmental, industrial, or educational employ are entitled to review and evaluate the work of other engineers when so required by their employment duties.
   c. Engineers in sales or industrial employ are entitled to make engineering comparisons of represented products with products of other suppliers.
8. Engineers shall accept personal responsibility for their professional activities, provided, however, that engineers may seek indemnification for services arising out of their practice for other than gross negligence, where the engineer's interests cannot otherwise be protected.
   a. Engineers shall conform with state registration laws in the practice of engineering.
   b. Engineers shall not use association with a nonengineer, a corporation, or partnership as a "cloak" for unethical acts.

9. Engineers shall give credit for engineering work to those to whom credit is due, and will recognize the proprietary interests of others.
   a. Engineers shall, whenever possible, name the person or persons who may be individually responsible for designs, inventions, writings, or other accomplishments.
   b. Engineers using designs supplied by a client recognize that the designs remain the property of the client and may not be duplicated by the engineer for others without express permission.
   c. Engineers, before undertaking work for others in connection with which the engineer may make improvements, plans, designs, inventions, or other records that may justify copyrights or patents, should enter into a positive agreement regarding ownership.
   d. Engineers' designs, data, records, and notes referring exclusively to an employer's work are the employer's property. The employer should indemnify the engineer for use of the information for any purpose other than the original purpose.
   e. Engineers shall continue their professional development throughout their careers and should keep current in their specialty fields by engaging in professional practice, participating in continuing education courses, reading in the technical literature, and attending professional meetings and seminars.

Footnote 1 "Sustainable development" is the challenge of meeting human needs for natural resources, industrial products, energy, food, transportation, shelter, and effective waste management while conserving and protecting environmental quality and the natural resource base essential for future development.

As Revised July 2007

"By order of the United States District Court for the District of Columbia, former Section 11(c) of the NSPE Code of Ethics prohibiting competitive bidding, and all policy statements, opinions, rulings or other guidelines interpreting its scope, have been rescinded as unlawfully interfering with the legal right of engineers, protected under the antitrust laws, to provide price information to prospective clients; accordingly, nothing contained in the NSPE Code of Ethics, policy statements, opinions, rulings or other guidelines prohibits the submission of price quotations or competitive bids for engineering services at any time or in any amount."

Statement by NSPE Executive Committee

In order to correct misunderstandings which have been indicated in some instances since the issuance of the Supreme Court decision and the entry of the Final Judgment, it is noted that in its decision of April 25, 1978, the Supreme Court of the United States declared: "The Sherman Act does not require competitive bidding."

It is further noted that as made clear in the Supreme Court decision:
1. Engineers and firms may individually refuse to bid for engineering services.
2. Clients are not required to seek bids for engineering services.
3. Federal, state, and local laws governing procedures to procure engineering services are not affected, and remain in full force and effect.
4. State societies and local chapters are free to actively and aggressively seek legislation for professional selection and negotiation procedures by public agencies.
5. State registration board rules of professional conduct, including rules prohibiting competitive bidding for engineering services, are not affected and remain in full force and effect. State registration boards with authority to adopt rules of professional conduct may adopt rules governing procedures to obtain engineering services.
6. As noted by the Supreme Court, "nothing in the judgment prevents NSPE and its members from attempting to influence governmental action . . ."

Note: In regard to the question of application of the Code to corporations vis-a-vis real persons, business form or type should not negate nor influence conformance of individuals to the Code. The Code deals with professional services, which services must be performed by real persons. Real persons in turn establish and implement policies within business structures. The Code is clearly written to apply to the Engineer, and it is incumbent on members of NSPE to endeavor to live up to its provisions. This applies to all pertinent sections of the Code.

**National Society of
Professional Engineers®**

1420 King Street
Alexandria, Virginia 22314-2794
703/684-2800 • Fax:703/836-4875
www.nspe.org
Publication date as revised: July 2007 • Publication #1102

INDEX

Notes:

Discussion questions are indicated by page numbers followed by parenthetical numbers, indicating the discussion question in which the topic may be found.

Case Studies and vehicle names (such as space shuttle *Challenger*) are indicated by italics.

Footnotes are indicated by page numbers followed by (f).

253